Springer Series in Design and Innovation

Volume 10

Springer Series in Design and Innovation (SSDI) publishes books on innovation and the latest developments in the fields of Product Design, Interior Design and Communication Design, with particular emphasis on technological and formal innovation, and on the application of digital technologies and new materials. The series explores all aspects of design, e.g. Human-Centered Design/User Experience, Service Design, and Design Thinking, which provide transversal and innovative approaches oriented on the involvement of people throughout the design development process. In addition, it covers emerging areas of research that may represent essential opportunities for economic and social development.

In fields ranging from the humanities to engineering and architecture, design is increasingly being recognized as a key means of bringing ideas to the market by transforming them into user-friendly and appealing products or services. Moreover, it provides a variety of methodologies, tools and techniques that can be used at different stages of the innovation process to enhance the value of new products and services.

The series' scope includes monographs, professional books, advanced textbooks, selected contributions from specialized conferences and workshops, and outstanding Ph.D. theses.

Keywords: Product and System Innovation; Product design; Interior design; Communication Design; Human-Centered Design/User Experience; Service Design; Design Thinking; Digital Innovation; Innovation of Materials.

How to submit proposals

Proposals must include: title, keywords, presentation (max 10,000 characters), table of contents, chapter abstracts, editors'/authors' CV.

In case of proceedings, chairmen/editors are requested to submit the link to conference website (incl. relevant information such as committee members, topics, key dates, keynote speakers, information about the reviewing process, etc.), and approx. number of papers.

Proposals must be sent to: series editor Prof. Francesca Tosi (francesca.tosi@unifi.it) and/or publishing editor Mr. Pierpaolo Riva (pierpaolo.riva@springer.com).

More information about this series at http://www.springer.com/series/16270

Nicola Morelli · Amalia de Götzen ·
Luca Simeone

Service Design Capabilities

Springer

Nicola Morelli
Aalborg University
Copenhagen, Denmark

Amalia de Götzen
Aalborg University
Copenhagen, Denmark

Luca Simeone
Aalborg University
Copenhagen, Denmark

ISSN 2661-8184 ISSN 2661-8192 (electronic)
Springer Series in Design and Innovation
ISBN 978-3-030-56281-6 ISBN 978-3-030-56282-3 (eBook)
https://doi.org/10.1007/978-3-030-56282-3

This Springer imprint is published by the registered company Springer Nature Switzerland AG
The registered company address is: Gewerbestrasse 11, 6330 Cham, Switzerland

Acknowledgements

This book would have never been possible without the contribution of the people that worked with the authors in the Service Design Lab at Aalborg University, Copenhagen. The authors' gratitude goes to Begüm Becermen, Kirsten Van Dam and Fanny Giordano, who read and discussed the content of the various chapters and provided valuable suggestions to improve it; to Giulia D'Ettole, who greatly helped with the visual design of diagrams and figures and with the final formatting of the book and who—together with Drude Holm Ehn—provided material for the 'Tryg Nørrebro Station' case; and to the students who read part of the manuscript (Troy Leininger, Tanner Slade, Hadas Zohar, Mirja Hopiavuori, Simona Mancusi) and provided useful feedback on its readability.

A special thanks goes to Language Editor Janet Feenstra, who critically proof-read the text.

This book is the result of many years' experience teaching and conducting research at the Department of Architecture, Design and Media Technology (Create) of Aalborg University. We thank the department for the funding it has provided and for its support of our activity, which has made this book possible.

We would also like to thank Misha Kaur, from the Australian Taxation Office, Giordana Ferri and, again, Giulia D'Ettole and Drude Emilie Ehn, for allowing us to publish some of the images in this book.

Last but not least, the book also originated within the context of the still ongoing EU-funded DESIGNSCAPES project, which aims at building the capacity for design-enabled innovation in urban environments. The book authors are grateful to all the DESIGNSCAPES partners for the many insightful discussions on design tools and methods.

Contents

About the Authors

Nicola Morelli is Professor in Service Design and Director of the Service Design Lab at Aalborg University. He has a Bachelor's in Architecture (University of Naples) and a Ph.D. in Industrial Design (Politecnico di Milano). His research focuses on methodological issues related to service design and on strategic issues related to the role of service design in social innovation, sustainability and socio-technical transitions. He participated in several European projects as main project coordinator, principal investigator and work package leader, for example, DESIGNSCAPES (H2020), MUV (H2020), Open4Citizens (H2020), EasyRights (H2020), My Neighbourhood (CIP-ICT-PSP) and Life 2.0 (CIP-ICT-PSP).

Amalia de Götzen is Associate Professor in Service Design and a member of the Service Design Lab at Aalborg University. She graduated in Electronic Engineering at the University of Padova and was awarded a Ph.D. in Computer Science from the University of Verona.

Her research activity focuses on digital social innovation, in particular, on the intersection between interaction design and service design. She is interested in tools and methods that bridge the analogue and digital world of services. She participated in several European projects as investigator and work package leader, for example, WeNet (H2020), Open4Citizens (H2020), DREAM (Culture 2007–2013) and ENACTIVE (FP6-IST), among others.

Luca Simeone is Assistant Professor in Service Design and a member of the Service Design Lab at Aalborg University. He holds a Ph.D. in Interaction Design from the University of Malmo. He is particularly interested in the organisational, economic and financial dimensions of design and on how design can support entrepreneurship and innovation processes. He participated in dozens of European projects as innovation manager and work package leader, for example, 3D Tune- In (H2020), PLUGGY (H2020) and Scent (H2020), among others.

List of Figures

List of Tables

Chapter 1
Introduction

Abstract This chapter introduces the core theme of the book. It explains why we decided to write it and discusses the concept of capabilities in relation to expert professional knowledge. We also provide a first definition of the core capabilities of service designers along with an overview of the structure of the book.

1.1 Why Are We Writing This Book?

Over the course of a long maturation period, the discipline of service design has seen the development of a methodological toolbox based on the tools used in design projects carried out by private companies, public organisations and within educational and academic environments. The many scholars working on service design and the wide range of disciplines involved in service design activities have made it possible to elaborate and define these tools, which have been specifically developed for the aims and needs of designers working in various phases of the design process.

This work has proven particularly useful because service design is intrinsically multi-disciplinary. The earliest application of the concept of designing a service originates from marketing studies (Shostack 1982). Later, this was adopted by other design disciplines, such as industrial design and interaction design, which, in parallel, contributed to the construction of an organic body of techniques to deal with different aspects of services. These include time-related and interaction aspects, which have certain characteristics that categorise them as processes rather than permanently defined products (Morelli 2002; Moggridge 2007; Bitner et al. 2008; Kimbell 2009; Löwgren and Stolterman 2004). Other aspects related to the negotiation of values between service providers and customers opened the perspective of services as a socially constructed activity (Morelli and Loi 2001), which suggested an exploration of methodological approaches of social theories that could help designers understand and manage the social and cultural aspects of services. Many studies have also explored contributions from engineering and production systems (Hollins and Hollins 1993), management (Normann 1991, 2001; Normann and Ramirez 1993), or service design's original field of marketing (Gronroos 1990). In addition, the growing relevance of technology infrastructure, such as online platforms and services, has pushed

N. Morelli et al., *Service Design Capabilities*, Springer Series in Design and Innovation 10, https://doi.org/10.1007/978-3-030-56282-3_1

designers to explore data as a new material for service design, evidencing the need for a revision of the tools that designers use through a more data-driven inquiry (de Götzen et al. 2018; Kun et al. 2019).

This intense and multifaceted activity has resulted in many online and offline manuals, toolkits, and textbooks aimed at supporting the work of service designers. These available resources are certainly a good sign of the increased interest in this discipline but may also be a source of confusion for those who have little experience and navigation skills. This is one reason why we decided to write this book. This wide range of tools can give the impression that they are *the* solution for designers and that simply using such tools implies a successful design process. However, knowing about certain tools or having used them a few times does not necessarily mean that one is an experienced service designer. For example, one can use pliers, a hammer and a saw yet be unable to refer to themselves as a plumber, electrician or carpenter. A plumber is a plumber because the person has expert knowledge of how to understand and work on a plumbing problem and will know the purpose of each tool and how to use it within specific contexts and circumstances. This expert knowledge forms the core of professional activity, where expert knowledge is even more crucial than the professional tools.

Expert knowledge is what professional experts need to navigate the possibilities their tools offer to find the correct sequence of actions that will lead to a solution to the problem at hand. Such knowledge is not implicit in the tools themselves but rather relates to previous professional experience, the capability to analyse and understand the problem, and interactions with the people and technologies that are part of the problem. For a plumber, what seems routine is in fact the synthesis of such knowledge and the result of the plumber's professional capabilities that we recognise when we call a plumber or when we pay the honorary for their work. Along the same line, this book aims to explore the body of service design-related knowledge to define the capabilities of designers and what they can offer as professionals.

However, when talking about service design, matters can get even more complicated because, while a plumber's expertise is more or less confined to a number of problems that concern water, pipes and related devices (which are the material of their work), designers, particularly service designers, struggle to define the material they work with (Blomkvist et al. 2016). Many service designers refer to the industrial design tradition, and in that context, the solidity of the material to be handled provides certainty and concreteness to the design profession. However, the extension of the notion of design on different areas of intervention (Buchanan 2001) and on services in particular has expanded the domain of expertise to extend beyond the material, and today, designers offer professional support in different areas ranging from healthcare and prevention services to policy making. Accordingly, the capabilities needed to deal with those problems are much wider and often require service designers to complement their work with capabilities from other disciplines when not directly collaborating with other experts. As a result, the core capabilities of service designers are becoming much harder to identify, and therefore, a new definition is needed.

Today, this need for a new definition is more relevant than ever because service design is forming part, or may even be the core, of specific educational programmes, and service designers are becoming increasingly needed in various areas of our social and economic systems. It is becoming ever more critical to make the profile of a service designer clear, for example, to students in an academic course or for a new position in a professional environment (Ehn et al. 2020). Several different professional profiles are emerging as a consequence of the complexification of societies and economic systems, and this may cause an overlap of different capabilities or create redundancy or friction in the collaboration among different capabilities in professional teams.

In the definition of the core capabilities of service designers and their area of expertise, it is important to pin down exactly what service designers can bring to their professional teams: what can they do? How can they help private and public organisations in their innovation processes? What are the capabilities they will 'sell' to their professional partners? What capabilities can be defined as 'the core' of the service designers' expertise? The definition of these capabilities is not only a common concern of service design educators but also a relevant matter for professional consultancies that integrate service design into their organisation, which is another reason why we felt it was important to write this book.

This book also gives us the opportunity to clarify designers' capabilities in relation to a new perspective that frames the activity of service designers in different areas and levels of intervention. Service design research is shifting the paradigm from a perspective that considers services in relation to goods (and their related production system) to a new perspective that involves different actors in a process of value co-creation. In the new perspective, designers not only design services but also work to facilitate the emergence of design capabilities that are latent in communities and individuals or are inherent properties of contexts (like cities or neighbourhoods). Service designers design with others, for services (or for value creation), and in different logical contexts. Thus, the final aim of this book is to revisit service designers' capabilities in light of the new roles that have opened up in innovation processes on different scales.

1.2 What Do We Mean by 'Service Design Capabilities'?

Publications in organisational studies, management, and human resources use terms such as 'capabilities', 'competencies', 'skills', 'talents' and 'ability', but in most cases, it is difficult to clearly distinguish between them (Teodorescu 2006; Acklin 2013; Delamare Le Deist and Winterton 2005). Of these, the term 'capabilities' has the broadest meaning, as it has been used extensively to describe both organisational capabilities (Teece et al. 1997; Zahra and George 2002; Barney 1991, 2001) and an individual capability, in which the latter is seen as involving the confidence to apply knowledge and skills within varied and changed situations (Stephenson and Weil 1992). The body of literature also considers capabilities as linked to the resources that an individual or an organisation has access to (e.g. financial resources, raw materials, machinery, software applications). In this book, we adopt a clear-cut characterisation borrowed from research in strategy that simply distinguishes between resources and capabilities:

> Resources are the assets that organisations have or can call upon and capabilities are the ways those assets are used or deployed. […] A shorthand way of thinking of this is that resources are 'what we have' (nouns) and capabilities are 'what we do well' (verbs). Other terms are sometimes used, for example, 'capabilities' and 'competences' are often used interchangeably. (Johnson et al. 2017, p. 80)

For example, a service designer may have access to resources like funding or a software application to create interactive visualisations or an innovation space where they can invite users to collaborative design sessions. But to use these resources wisely and effectively, the service designer should have the right capabilities: they must know how to make a financial plan for the project, how to use the software application to produce engaging and impactful visualisations, and how to smoothly facilitate a collaborative design process. This is what we refer to when we use the word 'capabilities', and this includes a broad spectrum of skills, talent, and specialised knowledge and abilities. In this book, we occasionally refer to competences so as not to overuse the word 'capabilities'; however, in line with Johnson and colleagues, we mainly use the two words interchangeably.

Within design research, many authors have studied capabilities in design (for some recent examples, see Lin 2014; Mortati et al. 2014; Manzini 2015; Wrigley 2016; Geraghty and Charnley 2016). This book builds on these contributions, particularly the work of Conley (2010). While academic literature has explored capabilities in relation to design more broadly, few studies are exclusively dedicated to examining capabilities in service design (Bailey 2012; Malmberg and Wetter-Edman 2016). This book aims to fill this gap.

1.3 Core Design Capabilities

When working on services, designers should apply a number of capabilities, some of which are typically personal (e.g. empathy, the capability to understand logical or social contexts), while others are generically professional (e.g. business capabilities, organisational capabilities, sensitivity to aesthetics and form). This book focuses on capabilities that specifically involve the design of services. Broadly speaking, these include the capabilities to inquire into context, provide perspectives on possible future situations, and structure design processes. To give an even clearer picture, we include a list of the specific capabilities in focus:

- *Addressing the context*: identifying and responding to relationships between a solution and its context
- *Controlling experiential aspects*: empathising with people and addressing experiential features of possible solutions

- *Modelling*: simulating, visualising and experimenting with possible solutions before all the information is available and using form to embody ideas and communicate values
- *Vision building*: imagining feasible, possible and desirable futures
- *Engaging stakeholders*: initiating and facilitating participatory co-creation processes
- *Working across different logical levels*: shifting from operative levels to different levels of abstraction
- *Building logical architecture*: articulating or identifying logical structures to frame problems and creative activities
- *Open problem solving*: identifying solutions across different logical domains and within uncertain and ambiguous contexts.

Capabilities, such as those listed above, generally refer to different strategies and actions according to different levels of intervention. The same capability will produce different effects and support different strategies depending on whether the designer is supporting people's interaction in the *value creation context* (see Sect. 2.2), designing the structure of a service, or contributing to policies or strategies that aim to change the institutional context.

1.4 The Structure of the Book

This book stems from the experience of the Service Design Lab at Aalborg University, an active research lab based in Copenhagen. Over the past two decades, the lab members have extensively studied and written about service design and directly worked on dozens of service design projects, both large and small. This book was conceived at the intersection of design research and practice and originates from our daily work and attempts to define, characterise, teach and apply service design capabilities with our students and partners. The book is structured into eight chapters:

Chapter 2 specifies the approach to service design that this book intends to propose and introduces the framework used in the chapters that follow to discuss service designers' capabilities. In particular, after an introduction on the nature of services and the evolution of the concept of value creation, three logical levels are presented in which design action is framed: one level that focuses on service as interaction, another level focusing on service as infrastructure, and a final level focusing on service as systemic institution.

Chapter 3 contains a synthetic description of eight core service design capabilities: addressing the context, controlling experiential aspects, modelling, vision building, engaging stakeholders, working across different logical levels, building logical architecture, and open problem solving.

Chapter 4 considers the service design capabilities required at the level of 'service as interaction'. This level concerns the precise moment in which value is created—the moment in which individuals, groups of people, citizens or service beneficiaries (we use different ways of indicating the main subjects, depending on context) interact with the service infrastructure, with peers, or with technological components, all with the aim of creating value. This is the crucial level of value creation, where the designers' capabilities complement and sometimes support the capabilities or knowledge of others. The capabilities outlined in this chapter indicate a specific design action in relation to the actions of other actors involved in the value co-creation system.

Chapter 5 considers the logical level of 'service as infrastructure'. This is the moment in which the service—in terms of its potential for value co-creation, or its value proposition—is defined through the appropriate ordering of human, organisational and technical factors. In other words, its infrastructure. This is the most familiar area for service designers, as it has been widely researched in the literature and in design education. This chapter highlights the capabilities designers use when designing services and clarifies the specific contributions designers can offer in collaboration with other professionals.

Chapter 6 focuses on the level of 'service as systemic institution'. At this level, the actions of designers are not aimed at defining changes or patterns of change but rather creating mechanisms for large-scale change. The role of designers at this level has not been sufficiently debated. Only in the last few years have designers started discussing their role in defining large-scale design and innovation changes. And despite these new discussions, the role remains unclear—both to other disciplines and the designers themselves. As a result, design action has not specifically referred to such changes except in a few cases where designers have participated in policy-making initiatives or analysed the scalability of local initiatives. Designers' capabilities to work and influence change on this level should be discussed in greater depth; therefore, these capabilities are the focus of this chapter.

Chapter 7 proposes the possible use of the conceptual framework outlined in the previous chapters to navigate the tools and methods available to service designers. This chapter is aimed at supporting designers or design students in building their own *operative paradigm*—that is, their own personal toolbox—on the basis of the levels of intervention and the capabilities they are recommended to use at each level.

Lastly, Chap. 8 presents some concluding remarks. A book can be thought of as a picture in time of a specific knowledge area, but knowledge by its own nature evolves continuously. This chapter offers some brief suggestions about what is beyond the frame of this picture and considers how the whole book can be used as a navigation tool to meet the present and future challenges facing service designers.

1.5 A Final Note About the Aims of This Book

After having outlined what this book is, it is important to explain what this book is not. It should be clear at this point that this book is not a collection of design tools for service designers. Given that the body of literature already proposes an exhaustive number of toolkits, this book aims instead to discuss service design capabilities in innovation processes, although any relevant tools are always mentioned in relation to specific design capabilities. We also provide references for and more information about each tool when relevant to specific design capabilities.

It should also be noted that, although this book does not intend to position service design within a philosophical and theoretical framework, it is nevertheless grounded in theoretical sources. The authors use these sources in their teaching and research, and it is through drawing from them that this book derives its logical structure.

References

Acklin C (2013) Design management absorption model: a framework to describe and measure the absorption process of design knowledge by SMEs with little or no prior design experience: design management absorption model. Creat Innov Manag 22(2):147–160. https://doi.org/10.1111/caim.12022

Bailey SG (2012) Embedding service design: the long and the short of it. In: Proceedings of the ServDes conference, Espoo, 8–10 Feb 2012

Barney J (1991) Firm resources and sustained competitive advantage. J Manag 17(1):99–120. https://doi.org/10.1177/014920639101700108

Barney J (2001) Resource-based theories of competitive advantage: a ten-year retrospective on the resource-based view. J Manag 27:643–650

Bitner MJ, Ostrom AL, Morgan FN (2008) Service blueprinting: a practical technique for service innovation. Calif Manag Rev 50(3):66–94

Blomkvist J, Clatworthy S, Holmlid S (2016) Ways of seeing the design material of service. In: Proceedings of the ServDes conference, Copenhagen, 24–26 May 2016

Buchanan R (2001) Design research and the new learning. Des Issues 17(4):3–23. https://doi.org/10.1162/07479360152681056

Conley C (2010) Leveraging design's core competencies. Des Manag Rev 15(3):45–51. https://doi.org/10.1111/j.1948-7169.2004.tb00171.x

de Götzen A, Kun P, Simeone L, Morelli N (2018) Making sense of data in a service design education. In: Proceedings of the ServDes conference, Milan, 18–20 June 2018

Delamare Le Deist F, Winterton J (2005) What is competence? Hum Resour Dev Int 8(1):27–46. https://doi.org/10.1080/1367886042000338227

Ehn DE, de Götzen A, Simeone L, Morelli N (2020) Do they (know they) need a service designer? An investigation of service design capabilities through the lens of the market. In: Proceedings of the ServDes conference, Melbourne, 2–5 Feb 2020

Geraghty RN, Charnley FJ (2016) Towards a design-in-government framework: a thematic analysis of case studies examining the capability of design in government. In: Proceedings of the 20th DMI: academic design management conference, Boston, 22–29 July 2016

Gronroos C (1990) Service management and marketing: managing the moments of truth in service competition. Lexington Books, Lexington

Hollins G, Hollins B (1993) Total design: managing the design process in the service sector. Pitman, London

Johnson G, Whittington R, Scholes K, Angwin D, Regnér P (2017) Fundamentals of strategy, 4th edn. Pearson Education, Harlow

Kimbell L (2009) The turn to service design. In: Julier G, Moor L (eds) Design and creativity: policy, management and practice. Berg, Oxford, pp 157–173

Kun P, Mulder I, de Götzen A, Kortuem G (2019) Creative data work in the design process. In: Proceedings of the ACM SIGCHI conference on creativity and cognition, San Diego, 23–26 June 2019 https://doi.org/10.1145/3325480.3325500

Lin JY (2014) Design capabilities in the public sector. In: Proceedings of the 19th DMI: academic design management conference, 2–4 Sept 2014. DMI, London
Löwgren J, Stolterman E (2004) Thoughtful interaction design: a design perspective on information technology. MIT Press, Cambridge
Malmberg L, Wetter-Edman K (2016) Design in public sector: exploring antecedents of sustained design capability. In: Proceedings of 20th DMI: academic design management conference inflection point: design research meets design practice, Boston, 22–29 July 2016
Manzini E (2015) Design, when everybody designs. MIT Press, Cambridge
Moggridge B (2007) Designing interactions. MIT Press, Cambridge
Morelli N (2002) Designing product/service systems: a methodological exploration. Des Issues 18(3):3–17. https://doi.org/10.1162/074793602320223253
Morelli N, Loi D (2001) Design and heterogeneous networks. A case study designing a service for internet access. In: First international conference on human aspects of the information society, Newcastle upon Tyne, 9–11 Apr 2001
Mortati P, Villari B, Maffei S (2014) Design capabilities for value creation. In: Proceedings of the 19th DMI: academic design management conference, DMI academic conference, London, 2–4 Sept 2014
Normann R (1991) Service management: strategy and leadership in service business. Wiley, New York
Normann R (2001) Reframing business: when the map changes the landscape. Wiley, Chichester
Normann R, Ramirez R (1993) From value chain to value constellation: designing interactive strategy. Harv Bus Rev 71(4):65–77
Shostack LG (1982) How to design a service. Eur J Mark 16(1):49–63
Stephenson J, Weil S (eds) (1992) Quality in learning. Kogan Page, London
Teece DJ, Pisano G, Shuen A (1997) Dynamic capabilities and strategic management. Strateg Manag J 18(7):509–533
Teodorescu T (2006) Competence versus competency: what is the difference? Perf Improv 45(10):27–30. https://doi.org/10.1002/pfi.4930451027
Wrigley C (2016) Design innovation catalysts: education and impact. She Ji 2(2):148–165. https://doi.org/10.1016/j.sheji.2016.10.001
Zahra SA, George G (2002) Absorptive capacity: a review, reconceptualization, and extension. Acad Manag Rev 27(2):185–203

Chapter 2
An Approach to Service Design

Abstract The approach to services has changed in the last decades: it has developed from viewing services in relation to their intrinsic differences in products to viewing services as processes of value co-creation. This chapter therefore introduces the evolution of this concept based on the early studies that introduced the idea that services could be designed to the latest approaches that frame services. It also includes a discussion of design action in relation to these at three logical levels.

2.1 Some Brief Historical Notes on the Idea of Service Design

Before defining our approach to service design, it is important to chart the evolution of this discipline from its origin to the approach we are proposing.

Services have existed since the earliest social aggregations of human beings, and they have always been *designed*—at least, in some way, in the form of organised labour (Blomberg and Darrah 2014; Kim 2018). The term *service design* emerged when the relevance of services in economic activities became evident and the need to properly organise the activities in a service emerged. The origin of the term hails from marketing literature. Shostack (1982), for instance, highlights the presence of service and product components in almost all market entities and the need to appropriately design all the components of a service. For this reason, she proposed the term *blueprinting* to describe the activity of designing and codifying the sequence of actions that are included in a service performance.

In the years that followed, service design was analysed from different disciplinary perspectives, unveiling specific relevant research areas in the field (Nisula 2012). Hollins and Hollins (1993), for instance, analyses services starting from an approach that focuses on the organisation of business operations and describes them as processes. In contrast, Mager (2008) focuses on the client perspective and on the interface between clients and service providers on the basis of which service solutions are to be visualised, formulated or orchestrated. Clatworthy (2010) proposes a similar perspective, which focuses on services as experiences that happen over time and that need to be organised through a sequence of interactions between service providers

N. Morelli et al., *Service Design Capabilities*, Springer Series in Design and Innovation 10, https://doi.org/10.1007/978-3-030-56282-3_2

and customers. In marketing studies, a similar perspective is centred around the *service encounter* (Czepiel et al. 1985), which is based on the interaction between customers and the service's tangible evidences (humans or artefacts) (Bitner et al. 1990). Service design has also been defined in relation to the coordination of the back stage of services, in other words, to the design of facilities, servers, equipment and other resources needed to produce services (Ghosh et al. 2004).

In those definitions of service design, two main directions emerge (Morelli 2009): whereas the first direction derives from the tradition of product and interaction design, which focuses on the front stage, on user experience and on the interface between service providers and customers, the second direction derives from management and marketing studies, where it focuses on services as processes and analyses the organisational aspects in the back stage.

To summarise the different contributions to the definition of service design, Kimbell (2011) highlights two main tensions: the first tension concerns understanding design either as a defined problem-solving activity or as an enquiry, which means an exploration of an open problem space involving different actors, including users. The second tension concerns understanding services on the basis of how they differ from products or as an activity of value creation.

The definitions derived from such tensions define design, either from an engineering perspective—keeping the distinction between products and services, and interpreting design as a problem-solving activity—or from a design-for-services perspective, which looks at services as a value creation activity in an open-ended problem exploration involving different actors (Fig. 2.1).

The two perspectives open different professional and disciplinary spaces: the former links service design to the tradition of engineering studies, in which design capabilities are derived by a broader definition of engineering capabilities, whereas

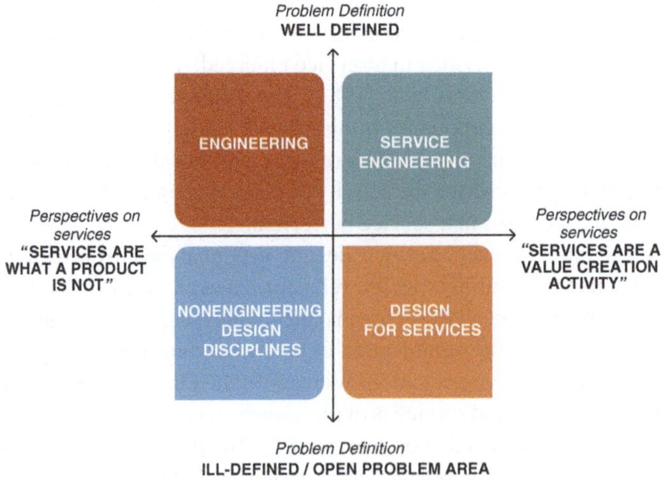

Fig. 2.1 Approaches to conceptualising service design. Adapted from Kimbell (2011)

the latter assembles knowledge domains that derive from different disciplines ranging from economic studies that focus on value creation to design studies that analyse the interaction among actors in a service system and all the way up to studies in the socio-cultural areas, which define the roles, knowledge and cultures that contribute to the shaping of services as a value creation process.

This book will mainly refer to the second perspective—focusing on value creation in an open-ended context—not only because of its proximity to design studies but also because the disciplinary convergence it implies requires the definition of the specific service design capabilities to be more articulated and tailored to each design action.

2.2 The Nature of Services in Relation to Value Creation

Common to the definitions analysed in the previous section is that they all start from the shared definition of a service as a fundamental activity in an economic exchange. In doing so, they take the etymological nature of the definition for granted. This book instead proposes to rediscover the original meaning of the term 'service', taking into account that this term can have different interpretations, and each interpretation can lead to different working areas for service design.

> A simple search of the term 'service' in the most common dictionaries provides different possible interpretations of the term.
> The first definition is provided by the *Oxford English Dictionary*: 'A service is the action of helping or doing work for someone.' The *Merriam-Webster Dictionary* defines 'service' as 'the occupation or function of serving', thus associating it with the activity of a servant. *Collins Dictionary's* first definition of 'service' is 'Something that the public needs, such as transport, communications facilities, hospitals, or energy supplies, which is provided in a planned and organized way by the government or an official body.' And our final definition from the most common dictionaries is given by the *Cambridge Dictionary*, where it instead defines a service as 'A government system or private organization that is responsible for a particular type of activity, or for providing a particular thing that people need.'

The etymological root of the word is the Latin term, *servus*, which means 'slave'. Therefore, a service is the work of a slave or a servant that offers their work for the benefit of another person (a master, a lord) with or without a monetary reward (Kim 2018). The various definitions given above refer to different interpretations of the term 'service':

1. Service as *interaction* between two or more people, characterised by unbalanced roles between *server(s)* and *served* (e.g. a nurse and a patient).
2. Service as an *infrastructure* that supports a certain kind of (service) activities (e.g. the hospital in which the interaction happens and its related organisation).
3. Service as a *systemic institution* (e.g. the institutional system of healthcare laws, the organisation of the healthcare system and the related scientific, technical and organisational knowledge) that organises the activities and processes.

The three concepts are equally important for the discipline of service design, as they represent three different working areas in which service designers operate. The development of capabilities for service designers must take into account these three areas.

It is worth noting that the common trait of the three interpretations is that all are centred around the process of creating value. The first definition observes the value creation process at the level of the interaction among the subjects that create value, the second at the level of the physical, functional or organisational infrastructure that makes the interaction possible, and the third definition focuses on the institutional level that represents the social, technical and regulatory context for the process of value creation. These three levels define different possible contexts for design. Taken together, these three points of observation define an ecosystem related to the production of value.

When I have a health problem, I may be able to solve it with my own knowledge (what I know about a headache or muscle pain) or my social knowledge (I can ask my trustworthy friends), or I can ask for help from a healthcare service (a doctor or a hospital). The process of value creation (healing my body) is therefore not necessarily using services. Whether I solve the problem myself or through a service, my action is framed by the institutional system of rules (there are drugs I cannot use, I may need to behave in a certain way to prevent infection, I must abide by general ethical principles), scientific knowledge (the scientific approach to the problem) and infrastructure (the way I can access healthcare services).

The interpretations therefore suggest that designing services means working on three logical levels, each of them requiring different capabilities (Fig. 2.2).

Before looking at such capabilities, it is worth focusing a bit longer on the definition of value creation and on the progressive evolution of the basic concept of value creation in the last few years.

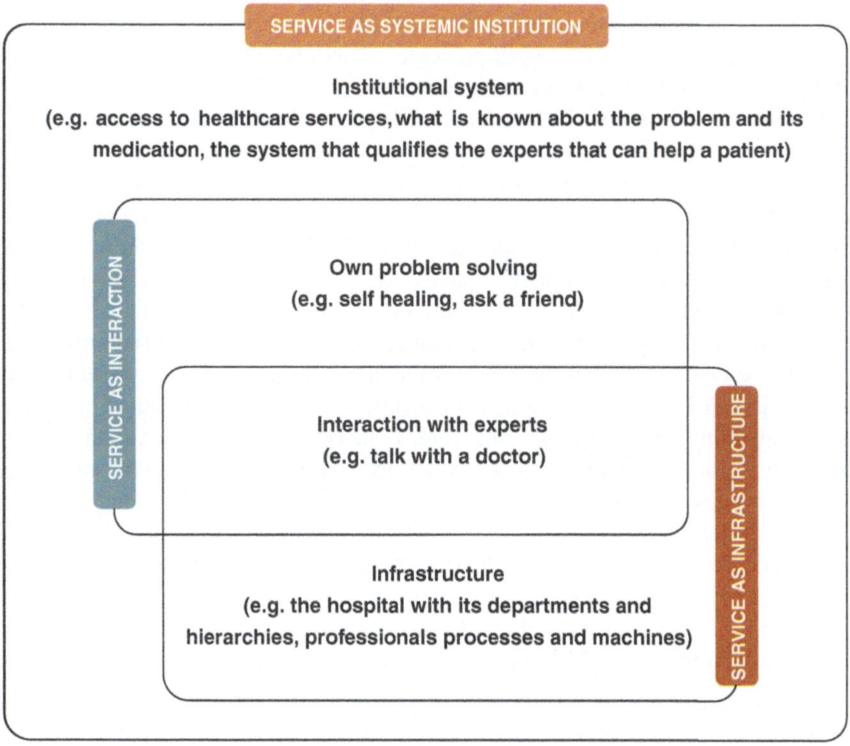

Fig. 2.2 The value creation ecosystem, from self-healing to healthcare institutions

2.3 Value Creation: The Evolution of a Concept

The concept of value and value creation has been debated since Aristotle (in Johnson 1939), but in the last few centuries, the theme of value creation has come to the fore and became the focus of much literature in economic studies. More recently, the significance of the value creation concept on the design discipline has been considered and its link between design and economics has been analysed.[1]

The traditional thinking about the process of the creation of value is based on a model in which a number of actors are aligned in a chain, which starts from upstream suppliers and continues by following additions down to the final customer (Johnson et al. 2017). Each actor in this chain provides input for the value creation of the actors downstream (Porter 1985; Normann and Ramirez 1993). In this perspective, the value creation process stops at the stage of the interaction with the final user. The image of the chain provides a simplified model of a system of value production. This model works well to represent some production processes, especially those

[1]Heskett et al. (2017) propose a comprehensive review of different economic theories and the role of design in creating value.

related to products but is less adequate for explaining more complex value production processes. The complexity of such processes cannot be interpreted through simple input–output or server–served mechanisms.

Normann and Ramirez (1994) propose a different picture to represent the process of value creation. They argue that 'the key to creating value is to coproduce offerings that mobilize customers' (p. 69). From this consideration, they outline a scenario in which the creation of value is the outcome of the interaction of a constellation of actors rather than a linear production chain. In this perspective, the role of the customer also changes from being a passive receiver of value in the chain model to becoming an active co-producer of value that interacts with other actors and also produces and aggregates resources (products, services, and infrastructures). This perspective is in line with what other authors have suggested. *Value creation* refers to the activity of creating something worthwhile, something we attribute importance to, or something deemed useful. Therefore, value is often measured in economics as utility. It is not an attribute of goods or services but rather linked to the subjective judgement of users (Heskett et al. 2017). Although the concept of value has often referred to qualitative criteria, such as pleasure and satisfaction, the need to deal with value in economics leads to the definition of value as something measurable and therefore related to an economic exchange. Furthermore, the concept of value has often been associated with the process of producing such value and embedding it into something that can be materially exchanged.

> Value for customers is created throughout the relationship by the customer, partly in interactions between the customer and the supplier or service provider. (…) The focus is not on products, but on the customers' value-creating processes where value emerges for customers and is perceived by them. (Grönroos 2007, p. 27)

> The customer becomes primarily an operant resource (co-producer) rather than an operand resource (target) and can be involved in the entire value and service chain. (Vargo and Lusch 2004, p. 11)

The value creation process, therefore, implies negotiation among different actors and may require facilitation in the form of interaction mechanisms. The context for this interaction is shaped by the infrastructure conditions (physical, functional or organisational) and the institutional conditions (cultural, political, social and economic frames) which facilitate, support or organise the value creation process.

A supermarket customer creates value by choosing the products needed to make meals (using their knowledge about the recipe). The infrastructure that facilitates this process (the supermarket) includes, for instance, the shelves, which organise and exhibit products, or even a combination of products that can be bought (e.g. using recipes) and the shopping cart. The institutional conditions refer to the implicit rules that customers should follow in the supermarket, in the payment arrangements (e.g. credit card system) and in the market rules that impose the price of the goods the customer intends to buy.

This perspective on the value creation process is changing the way in which services should be observed, and this book explores it as a new view of the service system. The exploration we propose stems from the interpretations of the dictionary definitions presented in the previous section and from the service design literature that looks at services as complex, nonlinear value creation processes. These interpretations help navigate different points of view on the value creation process, starting from the actual value creation moment and zooming out to see the bigger picture regarding the material and immaterial components of the whole service ecosystem.

Therefore, this book refers to these definitions in three different sections:

1. Service as interaction: facilitating the service beneficiaries in the time and context in which they interact with other actors and infrastructures to co-create value.
2. Service as infrastructure: designing the process and the place (the infrastructure) for value creation.
3. Service as a systemic institution: creating elements of changes (e.g. policies, scaled-up services) in the institutional system or aligning services to the institutional context, which includes the culture, social conditions, political frames and innovation attitude of a society.

The book contains subsections that briefly summarise the characteristics of these levels, while Chaps. 4, 5 and 6 give a more in-depth analysis of the levels in relation to the capabilities required for designers to operate at each level.

2.4 Service as Interaction: The Time and Context of Value Creation

The perspective at this level is based on an observation point that is very close in time and space to the interaction between two actors. As explained in Sect. 2.1, the etymology of the term 'service' and the definition provided by the *Oxford Dictionary* explicitly refer to an asymmetric interaction between two actors (or actor categories)—a *server* and a *served* actor. The definition assumes that there is an active actor who generates value and another actor who (more or less) passively benefits from such value. Early studies focused on services and service design (Shostack 1982, 1984) found that this assumption was highly efficient in describing a service, and in fact, many of the services that have existed since the beginning of our civilisation can be explained with this definition—from thermal baths in ancient civilisations to table service in a restaurant today.

This definition assumes that the value is entirely created by the *server*. The 'production process' of a service is therefore concluded in the time and place of the interaction between the server and the customer that takes benefit from the service (in this book, we indicate 'the customer' as the *beneficiary* of the service). In this sense, a service is not very different from a good, with the same asymmetric interaction mentioned earlier. As for products, the value creation process can also be considered as concluded at the point of sale when the good is handed over to the customer. In this moment there is a clear distinction between a *producer* and a *consumer*, and therefore, a clear distinction of their role: the producer is the actor (or the chain of actors) that has *stored* value in the product, whereas the consumer is the actor that *consumes* the value (the verb *consuming* derives from Latin, *consumare*, which means to finish, to conclude, but later also meant *to destroy*). According to this perspective the customer is not passive in the value creation process, but rather they are in fact *destroying* the value created by the product/service provider (Ramirez 1999).

A car or a piece of furniture is the output of a long chain of production processes—from the extraction of raw material, the production of components and the assembly, to the logistic processes that make them available at the selling point. But according to the value chain logic, right after these products are purchased and start to be consumed, their value is already substantially lower than the purchase price.

The analogy between services and products helped the early studies on services to qualify services as an economic activity: like products, services are produced by someone—they are part of an economic exchange between a producer and a consumer. They create value and are based on the work of someone for someone else. But the analogy was also used to define some basic differences between services and products that explain the main characteristics of services. Services were initially defined as 'that which is not a product', and therefore described as *immaterial*, *heterogeneous*, *inseparable* and *perishable* (Zeithaml et al. 1985)—from which, the acronym IHIP derives.

2.4.1 The IHIP Paradigm

For many years, the most common way to define the characteristics of services was based on the difference between products and services. The first difference between product and services is in the *immateriality* of a service. Products have a material consistency, whereas services are considered immaterial. An effective definition of services by the English magazine *The Economist* is that a '*service is everything that can't fall on your feet*' (Moritz 2009). In fact, this difference, which is very effective in describing one of the most evident characteristics of products, their material nature, no longer works when analysing the nature of services.[2] Services are the result of the combination of a number of factors, including humans, their knowledge, and the way humans experience services, but services also depend on material components, such as objects, technologies, and physical locations. Therefore, services do have a material component as well, but the emphasis on their immaterial values helps

[2]Today, the definition of products is becoming ambiguous, as IT experts would call a *product* a 'digital interface' and a bank clerk would use the same term to define a financial package.

to focus on aspects such as time, experience, and knowledge contribution from the various stakeholders, which are crucial for the process of value co-creation.

The core value provided by a school is immaterial: knowledge and the social experience of its students. Nevertheless, the school is usually a material place, with rooms, tables, chairs, teaching material and other material components, including the people that participate in this knowledge production process.

A second critical characteristic of services is *heterogeneity*. Heterogeneity describes the variability of services, both on the side of service providers and on the side of the customers: service change according to the input the customer puts in the service. As shown in the example below, the interaction between service providers and customers cannot be fully controlled by any of the actors because many factors emerge that depend on interpersonal relations between the service personnel and the customers.

A shoe store does not provide the same result to all its customers. The variability depends, among other factors, on the personal conditions of the shop attendant (how busy they are, how competent they are, how happy they are to do that job, or even how happy their life is in that moment) and on the characteristics of the customer (if they are looking for something special, if they have a precise idea of what to search for, how they pay, if they know the implicit rules of the shop, like waiting their turn or not touching the shoes on the shelves, or their personal characteristics, like their view on impulsive shopping).

Services' *inseparability* refers to how the production and consumption of services happen at the same time. While goods are first produced, then sold and consumed, services require the customer to be present in the very moment of value creation. This characteristic also refers to other issues concerning services, including the strong and intimate relationship the service provider and customer may have regarding the services.

While food preparation is an important function in a restaurant, the real value of the restaurant is created when the food is consumed (i.e. when the customer is in the restaurant). The experience a customer has in a restaurant highly depends on certain factors. These include the customer's emotional state and their dependence on other actors, for example, a waiter's capability to empathise and the possibility to engage in the restaurant experience by being informed about the ingredients or the preparation process of the food.

Finally, *perishability* refers to the fact that services cannot be stored or purchased (what is paid for is access to the service). This has implications on the synchronization of supply and demand. Services have to be available whenever the customer needs them and be able to adapt to the variation of the demand.

A gourmet restaurant cannot produce and store fully prepared meals in the precise moment customers arrive. The meals provided by the restaurant during peak hours have to be produced the very moment in which the customers visit the restaurant. This may create problems of limited capacity in high season.

2.4.2 The Limits of the IHIP Paradigm and Service Dominant Logic

The implication of the characteristics the IHIP paradigm refers to are synthesised by Zeithaml et al. (1985). Although the IHIP paradigm is very useful to describe some characteristics of services, it does not place enough emphasis on the nature of services as interactions between different actors. The exploration of such characteristics gives the opportunity to study services from a different perspective (Pacenti 1998).

Interaction aspects are not completely neglected in the IHIP paradigm, but they are not central. Their relevance also relates to the *time dimension*, which qualifies services as processes rather than products. Another consequence of this perspective is that the *value is co-created* in this interaction, instead of just being *passed* from a producer to a consumer. Services are an activity of value creation that continues (and sometimes becomes more intense) after the point of sale, for example, after the moment in which a contract for the service is signed or after the material resources to produce value are transferred from the service provider to the customer. The point of sale is the beginning of a co-creation process based on a collaboration between customers and other actors (including the service provider) (Fig. 2.3).

Normann and Ramirez (1993) emphasise how companies like IKEA have based their success on a concept of value co-creation that includes the customer as a main actor. The new business model proposed by the Swedish company was based on a new division of labour in which the customers implicitly agree to take over part of the tasks that traditional furniture companies cover. To do this, the company produced a number of *facilitation tools*, including (1) the catalogue, which is not just a collection of pictures of the furniture, but rather a *design manual* that suggests possible combinations of furniture to non-designer customers, of which, many would otherwise not be able to figure out all the possible solutions[3]; (2) the exhibition space—a 3D representation of possible configurations of the IKEA furniture; and

[3]The function of the catalogue in the IKEA strategy is further explained in Sect. 4.4.1.

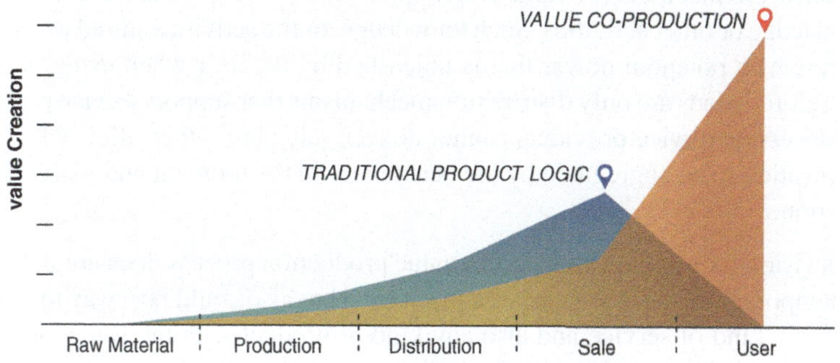

Fig. 2.3 Goods versus service value co-production. Inspired by Normann and Ramirez (1993)

(3) a number of facilities and infrastructures, from large parking areas to automobile roof racks to hire for transporting the furniture (Normann and Ramirez 1993).

Each product or service in the IKEA system is the result of a complicated set of activities. In fact, as Normann and Ramirez note, 'What we usually think of as products or services are really frozen activities, concrete manifestations of the relationship among actors in a value-creating system' (1993, p. 68). The actual production of value happens when the customer aggregates a number of resources (e.g. not only the IKEA furniture but also their own car to transport the items, their friends to help them in the transport, the mounting instructions) according to their own taste, knowledge, culture and personal capabilities. The value is in fact produced by a continuous interaction between the customer and a number of other actors, products and technologies. Normann and Ramirez observe that this happens not only with certain services but even when we use an ordinary product (Normann and Ramirez 1994).

A car is just a support for transportation—the real value is not in the car but on the constellation of products and services (e.g. roads, petrol stations, transport facilities) and in the knowledge (e.g. the driver knowledge about where to go and how to use a car) which aggregates the elements of such a constellation.

Seen in this perspective, the very moment of value production is in the interaction between the users/customers and the offering (products or services) that someone has sold or proposed to them. This perspective does not change whether we focus on a product (e.g. a car) or a service (e.g. IKEA) (Vargo and Lusch 2004, 2008). In both cases:

- Value is only produced when the customer comes into play. It is uniquely produced by the customer, when they aggregate a number of resources, which could be material (products) or immaterial (knowledge, concrete help provided by other people).

- Products are just tools for value creation, in which other people have frozen their knowledge or other activities. Such knowledge, or the activities stored in products, represents a potential power that is unleashed by the user when using products. Therefore, goods are only distribution mechanisms that support service provision.
- Producers or service providers cannot deliver value but rather offer only a value proposition to be aggregated by the beneficiary in the moment and place of value creation.

This vision focusing on service as a value production process does not only apply to contemporary services (e.g. service platforms) but also could be a way to read and analyse any kind of service and also any kind of product with the same logic. This is why Vargo and Lusch (2004, 2008) define services (*the application of specialised skills and knowledge*) as the fundamental unit of economic exchange.

The observation of services from a close perspective that focuses on interaction therefore makes it possible to define services according to a new logic—a *service dominant logic*. The definition of this logic as *service dominant* is in contrast with a *goods dominant* logic, in which (1) the purpose of economic activities is to make and distribute things (goods) that can be sold; therefore, goods (instead of services) are the main unit of economic exchange, (2) value is only produced and embedded in goods, and (3) users are *passive* in the process of value creation because they only use or consume the value embedded in a product or service (Table 2.1).

Table 2.1 Goods dominant logic versus service dominant logic

	Goods dominant logic	Service dominant logic
Primary unit of exchange	Goods	Services
Role of goods	Goods are end-products	Goods transmit embedded knowledge into the process of value creation
Role of customers	The customer is the recipient of goods	The customer is a co-producer of value
Value producer	Value is determined by the producer and embedded in goods	Value is perceived and determined by the customer
Firm/customer relation	Customers are passive	Customer is active, firms can only make a *value proposition* (through products/services)

Adapted from Vargo and Lusch (2004)

2.5 Service as Infrastructure: Designing the Process and Place for Value Creation

The perspective of this book is based on the assumption that design actions do not solely consist of complete beginning-to-end projects with a specified time span, and above all, with a specific outcome (value creation). If we assume that the aforementioned process of value creation depends on the presence and action of customers,[4] then all the actions *before* that moment—the work of service providers, the technical development of products and services, and the related design action—have to be considered as *propositions* rather than complete and stand-alone value-creation processes. In particular, the action of a designer has to be seen as an open-ended support to value creation. This perspective is in opposition to the view that sees the action of a service provider or a designer as a stand-alone project with a clearly defined outcome.

To mark the conceptual difference between a project-based approach and this open-ended approach, some authors (Björgvinsson et al. 2010; Hillgren et al. 2011; Karasti 2014) describe the design action as *infrastructuring*.

In a project-based perspective, designers create products or services and consider them as complete and stand-alone instances of value creation; therefore, their process has a precise end and is concluded when the product (e.g. a piece of furniture) leaves the manufacturer or when the service interaction with the customer is complete (with the final invoice). The result of such a process can be easily described by designers (e.g. through a blueprint),[5] who, together with service providers have full control over the possible configurations of the results of the process.

In the infrastructuring perspective, instead, the process of value creation starts during or after the intervention of a designer, and more specifically,

- through direct interaction between designers and customers in workshops and social innovation processes, or
- through material or immaterial service components that trigger customers' interaction with a service, such as service interfaces, prototypes, cards, or visualisations of the service offering.

In this perspective, the result of the process is open and cannot be completely described by designers (blueprints only describe possible use scenarios), and neither the designer nor the service provider have full control over the value produced.

---

[4]In this case, we are still referring to customers as the early literature on services does, but in fact, the actor playing this role in value creation has been defined as the *beneficiary* in the literature on the Service Dominant Logic. The literature on social innovation or on public services often refers to 'citizens' to indicate the same role.

[5]A blueprint is a technical drawing that specifies a design plan. In service design, the term is used to represent the sequence of actions in a service interaction, with a (more or less) detailed description of timings, actions, material evidence and actors involved in the interaction. Since the early contribution of Shostack (1982), who first introduced the idea that services should be designed, blueprinting has represented a key activity in service design.

Table 2.2 Project-based versus design-for-infrastructuring approaches

	Project-based approach	Infrastructuring approach
Process duration	Closed: the process has a beginning and an end	Open: the duration of the process depends on how the customer aggregates the resources at their disposal
Value created	Embedded in a product or service	Defined by the customer
Control over results	Designer/manufacturer/service provider controls the result	The customer controls the results
Description of the result (products)	Technical drawings/rendering/products	Scenarios of use catalogues/use proposition
Description of the result (service)	Blueprint	Scenarios of interaction/experience prototypes

The final value co-produced by the customer of a furniture shop is not in the furniture—which may convey a set of cultural or aesthetic values that will become part of the value co-creation process—but rather in the intimate relation the customer will establish with the space they can create through that furniture. This means that neither the designer nor the shop (the service provider) have full control of the value created with their support; they are just creating the infrastructure for the value creation process. Of course, designers and service providers control the production, distribution and sale of the furniture, but this does not fully encompass the value created (Table 2.2).

It is worth noting that the two approaches describe two different ways of looking at services. The choice or the preference for one or the other approach may depend on the business conditions or the nature of the service, but to a certain extent, any service can be observed from both perspectives; the choice of one or the other point of view will possibly highlight different aspects of the design action.

When designing services for a hospital, for instance, a project-based approach could be useful to design the platform that the patients will use to book an appointment, while an infrastructuring approach could be useful to design the whole healing process, in which the patient will interact with technologies, hospital procedures, and nurses or doctors.

2.6 Services as Systemic Institution: Introducing Elements of Change in the Institutional Context

So far we have analysed the activity of value co-creation and the role of designers in infrastructuring value co-creation, which means suggesting tools for activating customers' own cognitive and creative resources, together with proposing a context, a place, and a time sequence for value creation. Those activities, however, do not happen in a vacuum. They refer to the background context of laws, cultures, technical or social knowledge, regulations, physical and technical constraints. In other words, this background consists of the systemic institutions that shape our society. Such context links every design action to other actions that are being performed somewhere else, that have been performed before, or even that will be performed in the foreseeable future. This context is like a landscape—the scene on which human action is framed. It represents an exogenous environment beyond the direct influence of customers or firms.

This landscape also shapes innovation and changes at all logical levels, from individual choices to the design of complex service platforms or public institutions (Geels and Schot 2007). This context is relatively stable: changes in this landscape are usually slow and evolutionary, which means that any innovation is not the direct result of specific actions but rather the consequence of continuous negotiations among values, institutions and actors.

For this reason, talking about design in this context seems to make little sense. Design, seen as a purposeful action to change a present situation into a preferred one, has no chance to generate large systemic change. Yet, in the last decades, the emergence of large-scale problems and the acknowledgement of designers' responsibilities as part of those problems focused the attention of how design can have a role in systemic change. Bason (2014) argues that design is not traditionally associated with public policies, and asks, 'Can designers come to terms with the sheer scale, interdependence and complexity of public problems?' (p. 6). This is associated with the question of the extent to which design action can contribute to large-scale transitions. The question may become even more relevant with the recent emergence of the *mission-oriented innovation* approach (Mazzucato 2017), which sets broad and ambitious missions as a target for long-term policies and requires a large and collective design effort.

In this context, design can play a role in generating elements of change that have the potential to trigger larger systemic changes, for instance, by scaling-up local initiatives, thus working from a lower scale—a community or a small institution—to larger contexts, such as a city administration or national policies.

But a design approach can also be useful to align broad institutional changes to innovation in infrastructure and in smaller contexts, thus translating the potential of large policies into real changes in value creation processes.

In many countries, the welfare state that had been conceived and developed in the post-war period is experiencing a big crisis due to broad societal changes (Esping-Andersen 1996, 2002; Vandenbroucke 2003; Leadbeater and Cottam 2009). Several cases have been proposed for design interventions to create small local changes that could translate the need for a new welfare state into concrete cases by inspiring citizens to be activists or supporting their interaction with services and experts (Cottam and Leadbeater 2004; Manzini and Staszowski 2013). The challenge, however, remains with the issue of how such small change can, in turn, be scaled-up to re-create a big picture of the new welfare state that would mean translating those small and local innovations in terms that can trigger a new institutional system (Morelli 2015; Manzini and Rizzo 2011).

2.7 Summing Up: Working on Services on Three Logical Levels

This chapter introduced a logical framework for value creation that will be analysed in greater depth throughout the rest of the book. The framework is based on three logical levels:

- *Service as interaction*: At this level, value is co-created by service beneficiaries (customers) and other actors interacting with them.
- *Service as infrastructure*: At this level, services are organised as an open-ended support for the value co-creation process.
- *Service as a systemic institution*: This is the level in which broad cultural and social changes happen, and these frame the value co-creation processes and their related support infrastructure.

From the perspective of a designer, each of the logical levels listed in this chapter requires specific professional design capabilities for specific tasks. The chapters that follow will explore the way designers need to use their capabilities at each logical level. Such exploration is also meant to be a *navigation* tool that links service design tools (currently available in several publications) with the logical landscape in which service design will be used.

References

Bason C (2014) Design for policy. Routledge, New York

Bitner MJ, Booms BH, Tetreault MS (1990) The service encounter: diagnosing favorable and unfavorable incidents. J Mark 54(1):71–84. https://doi.org/10.2307/1252174

Björgvinsson E, Ehn P, Hillgren PA (2010) Participatory design and "democratizing innovation". Paper presented at the PDC2010, Sydney, 29 Nov–3 Dec 2010

Blomberg J, Darrah C (2014) Toward an anthropology of services. In: Proceedings of the ServDes conference, Lancaster, 9–11 Apr 2014

Clatworthy S (2010) Service innovation through touch-points: the AT-ONE touch-point cards. In: Proceedings of ServDes conference, Linköping, 1–3 Dec 2010

Cottam H, Leadbeater C (2004) Open welfare: designs on the public good. British Design Council, London

Czepiel JA, Solomon MR, Surprenant CF (eds) (1985) The service encounter: managing employee/customer interaction in service businesses. Lexington Books, Lexington

Esping-Andersen G (1996) Welfare states in transition: national adaptations in global economies. Sage, London

Esping-Andersen G (ed) (2002) Why we need a new welfare state. Oxford University Press, Oxford

Geels FW, Schot J (2007) Typology of sociotechnical transition pathways. Res Policy 36(3):399–417. https://doi.org/10.1016/j.respol.2007.01.003

Ghosh S, Surjadjaja H, Antony J (2004) Optimisation of the determinants of e-service operations. Bus Process Manag J 10(6):616–635. https://doi.org/10.1108/14637150410567848

Grönroos C (2007) Service management and marketing: customer management in service competition. Wiley, New York

Heskett J, Dilnot C, Boztepe S, Poggenpohl SH (2017) Design and the creation of value. Bloomsbury Academic, Oxford

Hillgren PA, Seravalli A, Emilson A (2011) Prototyping and infrastructuring in design for social innovation. CoDesign 7(3–4):169–183

Hollins G, Hollins B (1993) Total design: managing the design process in the service sector. Pitman, London

Johnson V (1939) Aristotle's theory of value. Am J Philol 60(4):445–451. https://doi.org/10.2307/290855

Johnson G, Whittington R, Scholes K, Angwin D, Regnér P (2017) Fundamentals of strategy, 4th edn. Pearson Education, Harlow

Karasti H (2014) Infrastructuring in participatory design. In: Proceedings of the 13th participatory design conference, Windhoek, 06–10 Oct 2014

Kim M (2018) An inquiry into the nature of service: a historical overview (part 1). Des Issues 34(2):31–47. https://doi.org/10.1162/DESI_a_00484

Kimbell L (2011) Designing for service as one way of designing services. Int J Des 5(2):41–52

Leadbeater C, Cottam H (2009) The user generated state: public services 2.0. https://www.partnerships.org.au/Library/Public_Services_2.0.htm. Accessed 28 June 2020

Mager B (2008) Design dictionary-service design. In: Erlhoff M, Marshall T (eds) Design dictionary: perspectives on design terminology. Birkhäuser, Basel, pp 354–357

Manzini E, Rizzo F (2011) Small projects/large changes: participatory design as an open participated process. CoDesign 7(3–4):199–215. https://doi.org/10.1080/15710882.2011.630472

Manzini E, Staszowski E (2013) Public and collaborative: exploring the intersection of design, social innovation and public policy. DESIS, p 181

Mazzucato M (2017) Mission-oriented innovation policy: challenges and opportunities. UCL Institute for innovation and public purpose working paper (2017-1)

Morelli N (2009) Beyond the experience. In search of an operative paradigm for the industrialisation of services. In: Proceedings of the ServDes conference, Oslo, 24–26 Nov 2009

Morelli N (2015) Challenges in designing and scaling up community services. Des J 18(2):269–290

Moritz S (2009) Service design: practical access to an evolving field. Stefan Moriz, London

Nisula JV (2012) Searching for definitions for service design—what do we mean with service design? In: Proceedings of ServDes conference, Espoo, 8–10 Feb 2012
Normann R, Ramirez R (1993) From value chain to value constellation: designing interactive strategy. Harv Bus Rev 71(4):65–77
Normann R, Ramirez R (1994) Designing interactive strategy. From value chain to value constellation. Wiley, New York
Pacenti E (1998) Il progetto dell'interazione dei servizi: un contributo al tema della progettazione dei servizi. Doctoral Thesis, Politecnico di Milano, Milano
Porter ME (1985) Competitive advantage: creating and sustaining superior performance. Free Press, New York
Ramirez R (1999) Value co-production: intellectual origins and implications for practice and research. Strateg Manag J 20(1):49–65
Shostack LG (1982) How to design a service. Eur J Mark 16(1):49–63
Shostack LG (1984) Design services that deliver. Harv Bus Rev 62:133–139
Vandenbroucke F (2003) Promoting active welfare states in the European Union. Lecture at the University of Wisconsin, Madison
Vargo SL, Lusch RF (2004) Evolving to a new dominant logic for marketing. J Mark 68(1):1–17
Vargo SL, Lusch RF (2008) Service-dominant logic: continuing the evolution. J Acad Mark Sci 36:1–10
Zeithaml VA, Parasuraman A, Berry LL (1985) Problems and strategies in services marketing. J Mark 49(2):33–46. https://doi.org/10.2307/1251563

Chapter 3
Core Service Design Capabilities

Abstract This short chapter provides descriptions of service designers' core capabilities. In the chapters that follow, these capabilities are discussed in relation to the three levels of design action.

The three levels for value creation mentioned in the previous session require a specification of the capabilities designers need in order to navigate and effectively work at each of those levels. The discourse surrounding the specificity of design capability is not new. Back in 1990, Cross highlighted the character of design abilities with the aim of enabling design educators to nurture such abilities in their students (Cross 1990). This is part of a wider debate on the foundation of design competences, started by the same author (Cross 1982) and aimed at qualifying design as a coherent discipline of study, clearly distinguished from areas related to scientific knowledge and humanities. The debate focused on the nature of design knowledge with the aim of defining the fundamental idea for a design culture (Nelson and Stolterman 2012). Although relevant for this debate, design practice was not its main focus because the aim of those studies was mainly to define the discipline. Design capabilities are instead more systematically investigated by Conley (2010), and in relation to service design, by Bailey (2012) and Malmberg and Wetter-Edman (2016). The original characterisation provided by these authors has been readapted in relation to the theoretical considerations presented in the previous chapter.

Addressing the context Every value co-creation action happens in a context, and it relates to the material and immaterial conditions of the context and the relation any solution can have within such a context. Analysing and understanding the context is usually the first action designers undertake in a design process. This analysis involves both the relationship between the individual and physical contexts, which concern individual attitudes, routines and behaviour. The analysis also examines the relationship between the individual and the socio-technical context, which focuses on social or technical knowledge, practices and shared problem-solving strategies.

Controlling experiential aspects This capability refers to the relationship between designers and those they work for. It concerns the knowledge designers have about the people involved in a service. Such knowledge usually goes beyond the mere

N. Morelli et al., *Service Design Capabilities*, Springer Series in Design and Innovation 10, https://doi.org/10.1007/978-3-030-56282-3_3

information about their needs or attitudes, implying that designers empathise with those people and deeply understand not only their routines, practices, attitudes and current needs but also the experiential quality a new solution would imply for them.

Modelling Modelling refers to designers' capability to simulate, visualise and experiment with possible solutions before all the information is available. Modelling and visualisation tools, such as visual representations or prototypes, are particularly powerful in the early phases of design processes when there are not enough elements to foresee a solution or even to frame the problem in a manageable way. The core capabilities of designers in this regard are not related to the fidelity of the model but rather to the quality of the knowledge a model can produce. For this reason, modelling can be used as an analytical tool (to create a model of a problem) or as a facilitation tool—a boundary object (Star and Griesemer 1989) that supports the interaction of key stakeholders in a service.

Vision building is a key characteristic of design activity. It consists of the capability to envision possible futures—however far in time and broad in scope they may be—and create consistent and structured visions of what that future could look like. The visions may consist of different kinds of representations, from simple narratives to complex scenarios, and from visual representations to simulations and prototypes.

Engaging stakeholders The process of value co-production is intrinsically based on the participation of an ecosystem of stakeholders. In particular, the service customer is always a value co-producer (Vargo and Lusch 2004, 2008). Designers have to be capable of identifying, mapping and engaging the stakeholders who will be part of the value co-creation process and to initiate and facilitate the process of value co-production. Such engagement will either lead to the direct involvement of some key stakeholders (e.g. the service beneficiary) in a co-design process for the definition of new solutions or to the creation of facilitation objects, mechanisms, and infrastructures that mediate and facilitate the process of value co-production.

Working across different levels of abstraction The operative context in which designers work often requires shrinking the focus of a designer's action to the minimum details of a service; however, actions and functionalities in services need to be codified and made accessible in different contexts and situations. This implies that designers should be able to take care of details and to abstract, which involves reducing the details to find broader frameworks to address the stakeholders' interaction in the service. This capability is also useful for understanding how even a minimum number of actions in a service can imply or generate broader systemic changes and institutional transformations.

Building logical architecture While the experience of a service may sometimes seem simple and linear, service design solutions often require that the organisation has complex logical and human technical and managerial structures. Such structures may be clearly visible in an organisational diagram but may also be articulated in a complex logical architecture, which also has institutional and socio-technical

Table 3.1 Descriptions of the design capabilities highlighted in this chapter

Name of the capability	General description
Addressing the context	The capability to identify and respond to relationships between a solution and its context
Controlling experiential aspects	The capability to empathise with people and address experiential features of possible solutions
Modelling	The capability to simulate, visualise and experiment with possible solutions before all the information is available, using form to embody ideas and communicate values
Vision building	The capability to figure out coherent possible futures
Engaging stakeholders	The capability to initiate and facilitate participatory co-creation processes
Working across levels	The capability to work through different levels of abstraction
Building logical architecture	The capability to articulate or identify logical structures to frame problems and creative activities
Open problem solving	The capability to identify solutions across different logical domains and within uncertain and ambiguous contexts

implications. The designer's capability in this respect consists of identifying the elements of such structures and finding ways for such elements to interact and work together.

Open problem solving As mentioned in the previous chapter, we are assuming that, in most cases, service designers operate in an open-ended context. As a result, contexts are weakly defined—with change outside the control of designers, or in fact, any individual actor. Instead, they are the result of interaction and negotiation among different actors. This requires the designers to conceive of solutions that need to be completed in the moment and context of value co-production. The output of designers' activity should be seen as a framework for possible solutions rather than a solution per se.

Table 3.1 summarises the capabilities described in this chapter. The logical levels outlined in the previous chapter involve different perspectives and challenge designers' capabilities in different ways. For instance, when working on services interaction, designers are required to use their capabilities to support or provoke value creation processes, and when organising services as infrastructure, designers use their capabilities to manage resources, knowledge and time sequences. Finally, when working at the institutional level, designers' capabilities are used to propose possible scenarios and link different changes. The chapters that follow will focus on the way designers' capabilities can be used at each logical level.

References

Bailey SG (2012) Embedding service design: the long and the short of it. In: Proceedings of ServDes conference, Espoo, 8–10 Feb 2012

Conley C (2010) Leveraging design's core competencies. Des Manag Rev 15(3):45–51. https://doi.org/10.1111/j.1948-7169.2004.tb00171.x

Cross N (1982) Designerly ways of knowing. Des Stud 3(4):221–227

Cross N (1990) The nature and nurture of design ability. Des Stud 11(3):127–140

Malmberg L, Wetter-Edman K (2016) Design in public sector: exploring antecedents of sustained design capability. In: Proceedings of 20th DMI: academic design management conference inflection point: design research meets design practice, Boston, 22–29 July 2016

Nelson HG, Stolterman E (2012) The design way: intentional change in an unpredictable world, 2nd edn. The MIT Press, Cambridge

Nielsen L (2013) Personas—user focused design. Springer, London

Star SL, Griesemer JR (1989) Institutional ecology, 'translations' and boundary objects: amateurs and professionals in Berkeley's museum of vertebrate zoology, 1907–39. Soc Stud Sci 19(3):387–420. https://doi.org/10.1177/030631289019003001

Vargo SL, Lusch RF (2004) Evolving to a new dominant logic for marketing. J Mark 68(1):1–17

Vargo SL, Lusch RF (2008) Service-dominant logic: continuing the evolution. J Acad Mark Sci 36:1–10

Chapter 4
Designing for the Level of 'Service as Interaction'

Abstract Value is co-created in the moment in which service beneficiaries interact with a service infrastructure, with other beneficiaries, or with technological components. The observation of services at this level reveals that different capabilities are involved in the process of value creation, from the everyday problem-solving capabilities of the service beneficiaries to the expert design capabilities of those who should facilitate value co-creation.

Two key terms have been used so far, which are very much related to each other and crucial for understanding the nature of the interaction that services imply:

- value co-creation, which indicates the very moment in which we produce value by accessing and/or interacting with a service (e.g. sending a message on a mobile phone, travelling on a bus, interacting with service personnel); and
- design, which indicates the act of planning a process of value creation.

Whenever we create value, we refer to a plan, which can either be implicit—when it refers to very well-known sequences of routinary actions, like preparing my breakfast—or it may need a structured design plan that requires certain design capabilities, like renovating the furniture in my kitchen. This means that value co-creation always implies a (implicit or explicit) design process.

The sections that follow focus on the process of value co-creation, the nature of the interactions this process implies, and the contribution of design (and expert designers) to this process.

4.1 What Is Design Action at This Level?

In the previous chapters, services are defined as interactions between multiple actors (also including non-animated actors such as objects and technologies) with the aim of producing value. This perspective offers a new point of view to interpret most of the actions in our individual life as a process of value creation.

N. Morelli et al., *Service Design Capabilities*, Springer Series in Design and Innovation 10, https://doi.org/10.1007/978-3-030-56282-3_4

Every day, in every moment, each of us performs actions, makes decisions and changes the existing course of things to adapt it to our material, ethical or spiritual needs. Whether we are preparing our breakfast or planning our trip to work, we are devising a course of action that is supposed to solve a problem or explore new opportunities for creating valuable results. We are creating value for our individual advantage (e.g. more comfort for ourselves) or for maintaining our social relationships (e.g. enhancing the interaction with those who live around us) or for taking care of our social environment (e.g. something that improves the quality of life in our community).

We create value by aggregating resources, such as the objects, technology, services, people or spaces around us. The knowledge of how to use these elements is also a resource we draw upon as we perform certain actions.

In most cases, our value-creation actions use the existing or 'conventional' knowledge that are implicit in our routines. This knowledge is the product of (even basic) technical know-how, such as how to use a coffee machine, together with knowledge derived from our social life, such as knowing where to buy the ingredients to make a meal. This know-how forms a body of skills (practice) that we can use at any moment. Of course, the actions involved in the decisions we make and the new value we produce are certainly not proposing anything particularly new. Those skills are individually developed in repetitive or recurrent activities (routines) and synthesised in standardised ways of performing a task. When performing such tasks, indeed, we refer to established protocols that also lower the effort of coordinating and aggregating the resources we have to perform any action or develop any solution.

We realise the relevance of such routines when we cannot use them. For instance, when we enter a supermarket we have never visited before and look for a product without knowing its location, we need to re-aggregate knowledge about product categories (e.g. fruit will be in the same area as vegetables). We also need to use the orientation skills we developed when visiting past supermarkets (e.g. pasta will be close to sauces) and our visual memory of product brands and packaging.

Such a routine represents a shortcut, without which, our daily life would be a nightmare of decisions; it would require continuous risk evaluation and constant assumption of responsibilities, even in actions that are repeated many times during the day.

Other actions in our daily life require that we use our critical sense (our judgement), our practical knowledge and our creative capabilities. These actions happen frequently in our daily life, such as when planning what to make for dinner or deciding what to do at the weekend. Most of those actions do not result in radical changes in our way of living or in our system of meanings, yet each of those actions generates value, which means it contributes to changing the existing situation to our preferred one.

Those actions can be defined as an activity of design, if we consider the definition of design proposed by Nobel laureate Herbert Simon: 'Everyone designs who devises courses of action aimed at changing existing situations into preferred ones' (Simon 1969, p. 55).

The activity of planning those actions, imagining the result of them, and figuring out how to organise the resources available to achieve the desired change, however implicit or routinary, can be qualified as a design activity.

The design capabilities we need to perform such activities are quite basic. They correspond to our problem-solving skills; therefore, if we accept the basic implications of Simon's definition, design can be defined as a *diffuse* capability (Manzini 2015), or as Heskett defines it, 'Design is one of the basic characteristics of what it is to be human and an essential determinant of the quality of life. It affects every aspect of daily life' (Heskett 2002, p. 4).

4.2 The Role of Design at This Level

When we look at service design as a form of interaction and value co-creation and observe it closely, we understand that the role of service beneficiaries is highly relevant. We also understand that they do not need to be professional designers to be part of this process because this interaction involves actions they have to perform on the basis of their own problem-solving capability.

As mentioned in the previous section, everyday problem solving refers to one's attitude towards solving recursive or trivial problems—like driving a car, cooking or going to work—which usually do not represent any particular challenge to one's individual problem-solving capability. This routine type of behaviour has been described and studied as *practice* in Social Practice Theory (Reckwitz 2002). *Practices* are defined as behaviours that depend on the interconnectedness of contextual elements, such as material objects and their use, background knowledge, know-how and emotional states (Reckwitz 2002, pp. 49–50). Practices are therefore a logical framework, and they define a *comfort zone* for our daily actions. Although such practices are constantly repeating sequences of actions, they may tolerate changes that exercise higher or lower pressure on the practice framework.

Although working within the boundary of everyday practice requires little specialised knowledge about design, performing actions beyond such everyday practice—beyond our everyday comfort zone—recurs in modern life, and it challenges our design (or problem-solving) capabilities. Almost every day, we are challenged by problems we have not met before, services we do not know, or procedures that we need to understand. Creating value (e.g. solving such problems, accessing such services, activating new processes) in those instances is not easy. Value, in this case, is created with the help of some type of support (mobile applications, communication, handbooks), which activates or stimulates customers' capabilities. The definition of such design support needs *expert* design capabilities.

Our daily trip to work is a routinary task that does not need to be overly planned, but reaching a place in a new city that is unfamiliar to us requires some 'expert' support. We can use our social capabilities—and ask around about how to get to our goal, assuming that others will know the place better than us and speak our own language—or we can use a map on paper or our mobile. Mobile maps in particular provide detailed and real-time suggestions about the best itinerary, with trip timings and any points of interest along the way.

Designing (i.e. generating purposeful change) at this level requires capabilities to interact with other actors or technologies or to organise such interaction, as in the organisation of information in a service activity.

Creating a new community of people living together (a co-housing community) may require an interaction that challenges individual capabilities to figure out the quality of the collaboration among citizens and of the services that are needed for a harmonic life together. *Fondazione Housing Sociale* (Social Housing Foundation) developed a set of cards (Fig. 4.1) that support communities by suggesting services that cover the most recurrent aspects of living together. Such cards facilitate the dialogue among the members of the community (Ferri 2016).

Fig. 4.1 Social cards: a card set that facilitates the identification of collaborative services that address everyday life occurrences in a social housing community (Ferri 2016)

When working at this level, it is worth noting that (a) none of the actors interacting for value creation has full control of the value-creation process and (b) physical technological, logical or organisational structures—such as cards, phones, procedures, shopping centers or websites—are *interaction devices* that can be designed to support value creation; they do not imply value per se but rather mediate between actors and facilitate the development of value in a specific time and context. It follows that at this level, designers and service providers are not *designing* services (i.e. the value creation process) but rather a number of *interaction devices* consisting of physical, technological, logical or organisational micro-structures that could facilitate the process of value creation.

According to this view, a furniture shop does not produce value in its furniture, or with its exhibition spaces, or when delivering and mounting the furniture, but rather the value is produced by its customers in the moment the furniture is used to improve the value of one's home life (Normann and Ramirez 1993).

Do expert designers have a role in the process of value co-creation? If we consider value co-creation as a process related to the very moment a product is used or a service is accessed and thus see the value as solely produced by the interaction between users/customers (with their own problem-solving capabilities) and the elements of their physical/logical environment, then the obvious question is whether design capabilities can be related to this moment, and if expert designers have a role in this process.

There is no doubt that the technical, physical and logical environments in which the interaction happens have been designed to a larger extent. The objects in the interaction, the sequence of actions and the role of people and objects involved in a service interaction are planned to suggest a certain kind of relationship. The quality and characteristics of those elements of a service suggest or make certain actions possible, where they implicitly or explicitly prevent other uses or behaviours. Who designs such elements cannot directly or exactly prescribe how the service will develop, but they envision the possible relationship between the properties of those elements and the capabilities of the customer. In Norman's terms (Norman 1998), they design the *affordance* of such elements.

4.3 Design Capabilities at This Level

The design action at this level should aim at facilitating the interaction between individual beneficiaries and the social and physical context in which their value co-creation action is placed. Therefore, framing and contextualising service interactions and supporting shared models of innovation or shared visions is useful. Designers can have a facilitation role if they are supported by the certain capabilities:

- The capability to empathise with people and address experiential features (visual, sensorial) of possible solutions (*Controlling experiential aspects*)
- The capability to facilitate the process of value creation (*Engaging stakeholders*)
- The capability to simulate, visualise and experiment with possible solutions before all the information is available, using form to embody ideas and communicate values (*Modelling*)
- The capability to identify and respond to relationships between a solution and its context (*Addressing the context*)
- The capability to figure out coherent possible futures (*Vision building*).

These capabilities are defined in more detail in the next section.

4.3.1 Controlling Experiential Aspects

Everyday value creation depends on the intersection of experiences, knowledge and practices from a variety of subjects: individuals, service providers, technological infrastructure (that brings about the technical knowledge and experience of those who designed it)[1] and other social actors.

The contribution that expert designers can offer to the process of value creation depends on the nature of the knowledge that designers want to gain about the people interacting in the value creation process and on the strategy they want to use to *activate* such knowledge. When designers do not have direct interaction with the actors in the value co-creation process, their strategy could consist of *embedding* the knowledge needed to use the service into objects, technologies, and services that can be proposed to the users. This is the case with the most common services, from supermarkets to ATM machines, which mediate the interaction between designers and customers. In those cases, given that the interaction between designers and the service customers is not direct, designers have to investigate the needs, practices and experiential knowledge that come into play. They have to empathise with the main actors contributing to the value creation processes. Design culture, particularly in the last few decades, has placed an intense focus on techniques to obtain user-related information by building upon collaboration with experts in ethnography, anthropology, social studies and interaction design.

[1]Every product, service or technical infrastructure embeds the knowledge of its creator. It gives the user a number of use options (affordances) and imposes limitations in regards to certain other use options. The concept of affordance is explained by Krippendorff (1989). A simple internet connection in a school, for instance, is (supposed to be) perfectly fit for the needs of the teachers but would impose big limitations for other uses, such as supporting the community outside of school hours (Morelli and Loi 2002).

Fig. 4.2 Services to support older people's independent life, based on a dedicated app for the older population to access a local social platform (Life 2.0 project)

When the interaction between designers and the other actors is direct, designers can propose prototypes or experiments that highlight the experiential aspects of possible solutions.

4.3.2 Engaging Stakeholders

When designers interact directly with the value co-creation system, their strategy to contribute to value creation may consist of directly activating the knowledge of service beneficiaries through direct involvement. This is the case with participatory actions, or design activism, in which the service is developed through the interaction with the service beneficiary. In this case, the service beneficiaries' capabilities, needs, and problem-solving attitudes emerge from their direct contact with designers rather than from an accurate user analysis. This strategy can be supported by *generative* or *facilitation tools*, such as cards (Murray et al. 2010; Morelli et al. 2017; Ferri 2016), gamification tools (Di Dio et al. 2019; Oliveira and Petersen 2014), design experiments (Cantu and Selloni 2013; Cearreta Innocenti et al. 2018) and apps, which spark conversation and collaboration between actors with different types of expertise, knowledge and culture (Fig. 4.2).

4.3.3 Modelling

Making solutions visible before all the information is available is a critical function of designers, if, as Simon suggests, design is about envisioning possible future conditions to improve those of the present. The *modelisation* of possible realities is a way to show how it is possible to change the present into a desirable future. Models of a future reality can take many forms, from bi-dimensional representation such as video (Vistisen 2016; Bolvig Poulsen and Morelli 2010) to 'pretotypes' (Savoia 2011) to drama (enacting scenarios) or more complex service prototypes in which different elements of a future solution can be simulated and tested.

The activity of design on this level may use tools, such as visualisation (journeys or storyboards), analytical tools (personas, experience-, context- or technical analyses) and models (prototypes, cards or role-playing), which can be used to propose changes, simulate and experiment new solutions, and figure out the role of each stakeholder in the value creation context.

4.3.4 Addressing the Context

The act of value co-creation happens in a context, and it is shaped not only by the stakeholders involved in the co-creation but also by the technological infrastructure and by the organisational, cultural, economic and natural conditions. The complex interaction of such elements makes the results of every value co-creation act different. In this perspective, it is important to identify the elements of the context, their interaction and the way they might react to a proposed change.

This capability requires an attitude that is open to recognising and mapping the ecosystem around each value co-creation action in order to understand and highlight their motivation. When working directly with the beneficiary (e.g. through design activism, workshops, co-creation sessions or hackathons), designers need to visualise and prototype to create realistic scenarios that represent how the change will impact the life, routines or business activities of the beneficiaries and other stakeholders.

4.3.5 Vision Building

Vision building is a typical capability related to designers' attitudes towards creative and exploratory processes. When designing to support value co-creation, vision building can help other stakeholders figuring out possible ways to put together different elements in a new ecosystemic configuration. Prototypes or narrative techniques move the observation point and the analysis of possible futures very close to the level of the interaction, and they focus on human or social aspects, thus helping the various stakeholders figuring out not only the functional components of the service but also any emotional aspects that would otherwise be left out of a more synthetic view of possible solutions.

Table 4.1 Practice and design capabilities at the level of value co-creation

Objects of the design activity	Characteristics of design activity	Examples	Capabilities required
Non-routinary problems	Interaction within communities	Solidarity purchasing groups, urban gardening groups. Social housing	Addressing the context Vision building Engaging stakeholders
	Interaction with individuals (directly or through technology)	Mobile/web application, interaction cards, games	Modelling Controlling experiential aspects
Everyday practice	Interaction with services	Eating in a restaurant	Personal experience
	Interaction with products/technologies	Driving a car	Personal problem-solving strategies

4.3.6 *Summary*

Table 4.1 presents some examples of problem solving and the design capabilities required at the time and place of value creation along with the most common tools to support such capabilities. The table includes activities that do not go beyond people's everyday practice, experience and problem-solving attitudes—and therefore can be performed without any special planning skills or design tools. It also includes activities that challenge such practice (non-routinary problems) and implies a purposeful change in people's routine activities, which means they may need the support of expert design capability.

4.4 Cases

4.4.1 *The IKEA Catalogue*

Introduction We are probably all familiar with IKEA's offerings, particularly in relation to ready-to-assemble furniture. As the world's largest furniture retailer, IKEA operates more than 400 stores in some 50 countries.[2] The furniture is generally sold in packages that can be easily brought home and is built by the customers following relatively straightforward instructions.

The success of the company is grounded on the idea of shifting certain tasks (logistic, transport) from the company to the customers. One important function that

[2]Source: https://www.ikea.com/. Accessed 4 March 2020.

the IKEA business model passes to the customers is design: while other more traditional furniture shops also provide design consultancies to their customers, IKEA's customers are challenged to figure out how their home will look once the furniture is in place or even what kind of living space they want to live in.

A core element of IKEA's strategy is their catalogue. In it, they present not only their products (e.g. tables, chairs, couches) but also a repertoire of pictures that show how the products will look in situ, for example, how a certain table will look when put close to a coloured wall or cupboard and what kind of living space they can create by putting together different items. Flipping through the pages of an IKEA catalogue, one can see typical scenes like that of a family enjoying a meal in their dining room, a dad reading a book to his daughter in her bedroom, and a woman working at her desk in her studio. In each catalogue, dozens of similar photos show how the IKEA furniture and home accessories can be combined to create environments that are well styled, cozy, creative, functional and welcoming. Every year, about 200 million IKEA catalogues are printed and distributed.[3]

This catalogue is a core element, as it provides the customers with a repertoire of possible IKEA-powered solutions to furnish and decorate their home and workplaces. In fact, the catalogue is a design support, in the sense that it invites the customers to use, re-interpret, appropriate, adapt, and combine IKEA furniture and home decorations and act as interior designers. In other words, the IKEA catalogue is a device that inspires the customer with how their interaction with the service can lead to value production (Fawzy 2019).

Role and challenge for designers Of all the services offered by IKEA, the catalogue is one of the elements that more fully invites and guides customers in processes of value co-creation. In most countries, IKEA sends these catalogues out through the post. When IKEA designers work on the yearly catalogue, they provide suggestions, examples, instructions, and a creative repertoire that takes the customers by the hand and shows them a world of possibilities. Not all of us have the capability to envision a space that does not yet exist or the visual effect of different colours or material in a space. Not everyone has this creative capability, not to mention the capability to figure out the size of a piece of furniture in relation to the physical dimensions of a space. In this case IKEA provides a kind of 'design manual' in which customers can imagine their life and sometimes recognise themselves in the spaces represented. With this support, and with a visit to the physical space where they can see and touch the items they are going to buy, customers are much more confident to work on their own (without the help of a professional interior design consultant) in the creation of their own living space.

Design capabilities involved In this case, certain design capabilities played an important role:

- *Controlling experiential aspects*: while working on the catalogue, the IKEA designers have to empathise with people and imagine an array of possible solutions

[3]https://www.statista.com/statistics/268131/number-of-printed-ikea-catalogs-per-year-worldwide/.

based on the very different needs, interests and lifestyles of customers throughout over 50 countries and in diverse geographic, social and cultural contexts.

- *Engaging stakeholders*: the powerful suggestions of the photos in the catalogues encourage the customers to engage in the design process, as the catalogue not only shows the aggregation of furniture in a living space but also makes clear how possible combinations can be created through different aggregations of modular furniture.
- *Vision building*: the catalogue is clearly a source of inspiration for customers. It requires a capability of aggregating different configurations of furniture according to visions of different living spaces. The customer has to be able to envision themselves and their life in the space.
- *Modelling*: this capability refers to the creation of prototypes represented in the catalogue or directly in the physical stores, which propose different experiences to the customers.

4.4.2 Tryg Nørrebro Station

Introduction Public space is per definition something that belongs to everyone. However, not everyone has the possibility to decide how public spaces should be designed. When involving the public in planning projects, engagement practices are often reduced to citizen hearings that only include and attract a specific part of the population.

With the aim of making citizens an active part in the decision-making process of urban development, a team of designers based in Copenhagen decided to involve a citizens' alliance (*Medborgerne*) in the activation of temporary urban common spaces through a bottom-up design approach. A local union was interested in drawing political attention to and making unified demands to improve a local train station in the district of Nørrebro. The space in question had been under construction for a vast period of time due to a new metro line, and the degradation of the space had driven a part of the local community to feel unsafe in the area.

Despite Medborgerne's interest in improving the area, no physical action was taken. They encountered both a lack of established practices and processes on how to intervene about the space and a lack of clear self-governance regarding the urban resources.

A design team formed by a group of students decided to assist the local community in influencing the future of the area around the station through a series of design experiments. The aim of the team was to improve the livability of the public space by strengthening a collective sense of ownership in the space. This, in turn, would support the co-creation of value in the form of social capital.

The team organized a co-design workshop with Medborgerne's community to brainstorm on possible design experiments—meaning temporary urban activities—that would improve or question the liveability of the place during a specific period of time. The goal of the workshop was to get as many people involved in the design process as possible so that they could feel a sense of ownership and more strongly bond with the area.

The design experiments were located in various spots around the station area in order to activate and question the use of the space and to ensure visibility to as many people as possible. Furthermore, the experiments acted as 'interaction devices' designed to engage people in the conversation about the future of the area, enabling them to share ideas and visions and co-create possible solutions.

Role and challenge for designers The role of the designers was, first of all, to uncover and convey divergent interests concerning the area so that different actors could be empowered when involved in the workshop and in the participatory design experiments.

The designers gathered and communicated ideas (from and to the participants) by providing different interaction devices to boost engagement, facilitate co-creation, and collect insights. The interaction devices had different formats—ranging from a common diary to letters, posters, graffiti, a canvas for sketching and a temporary public living room—and different purposes. All the devices designed by the team acted as a means to trigger/provoke, as they encouraged citizens to interact and empathise with the space, exchange perceptions with their neighbours, and imagine future development scenarios for the area (Fig. 4.3).

Moreover, by translating insights into actionable points, the designers acted as mediators between citizens, community spokespersons and decision makers.

Design capabilities involved In this case, specific design capabilities played important roles:

Fig. 4.3 Pictures from the design experiments proposed to the community living around Nørrebro station. *Source* Cearreta Innocenti et al. (2018)

- *Engaging stakeholders*: The experiments proposed by the design team had the scope of triggering stakeholders' participation, creating a place for public conversation and supporting the interaction of the local community with their environment, in a place that was otherwise neglected by citizens.
- *Addressing the context*: Before intervening in the public space, the designers carried out a long process of identifying the neighbourhood and its prominent community voices and the space itself. The diversity of voices and visions in the area was reflected in the diversity of tools and activities that were designed.
- *Vision building*: By providing a blanked-out map of the area for sketching future scenarios, it was possible to trigger the citizens' participation while also providing a confined space for ideation. This allowed for not only involving a diverse group of citizens but also guiding their ideas toward more tangible solutions.
- *Modelling*: By adding living room furniture to the public space, a prototype of how the public space could be appropriated was suggested, proposing a different kind of experience to citizens.

References

Bolvig Poulsen S, Morelli N (2010) From research-based teaching to teaching-based research and back. In: Proceedings of the international conference on engineering and product design education, Norwegian University of Science and Technology, Trondheim, 2–3 Sept 2010

Cantu D, Selloni D (2013) From engaging to empowering people: a set of co-design experiments with a service design perspective. In: Social frontiers: the next edge of social innovation research. NESTA. https://www.scribd.com/document/191799251/From-engaging-to-empowering-people-a-set-of-co-design-experiments-with-a-service-design-perspective. Accessed 25 June 2020

Cearreta Innocenti T, D'Ettole G, Holm Ehn DE (2018) Supporting the activation of common spaces: a shift from consumers to co-producers of urban public spaces. Thesis dissertation, Aalborg University, Copenhagen

Di Dio S, Lissandrello E, Schillaci D, Caroleo B, Vesco A, D'Hespeel I (2019) MUV: a game to encourage sustainable mobility habits. In: Gentile M, Allegra M, Söbke H (eds) Games and learning alliance. GALA 2018. Lecture notes in computer science, vol 11385. Springer, Cham pp 60–70. https://doi.org/10.1007/978-3-030-11548-7_6

Fawzy RM (2019) Neoliberalism in your living room: a spatial cognitive reading of home design in IKEA catalogue. Discourse Context Media 31. https://doi.org/10.1016/j.dcm.2019.100309

Ferri G (ed) (2016) Starting up communities. Un Design-Kit per l'Abitare Collaborativo. Mondadori, Milan

Heskett J (2002) Toothpicks and logos: design in everyday life. Oxford University Press, Oxford

Krippendorff K (1989) On the essential contexts of artifacts or on the proposition that "design is making sense (of things)". Des Issues 5(2):9–39

Manzini E (2015) Design, when everybody designs. MIT Press, Cambridge

Morelli N, Loi D (2002) Design and heterogeneous networks. A case study designing a service for internet access. In: Proceedings of the first international conference on human aspects of the information society, Newcastle upon Tyne, 9–11 Apr 2002

Morelli N, Aguilar M, Concilio G, de Götzen A, Mulder I, Pedersen J, Torntoft LK (2017) Framing design to support social innovation. Des J 20(Sup 1):171–184

Murray R, Caulier-Grice J, Mulgan G (2010) The open book of social innovation. In: Social innovator series: ways to design, develop and grow social innovation. Young Foundation, NESTA, London
Norman DA (1998) The design of everyday things. MIT Press, Cambridge
Normann R, Ramirez R (1993) From value chain to value constellation: designing interactive strategy. Harv Bus Rev 71(4):65–77
Oliveira M, Petersen S (2014) The choice of serious games and gamification. In: Ma M, Oliveira MF, Baalsrud Hauge J (eds) Serious games development and applications. SGDA 2014. Lecture notes in computer science, vol 8778. Springer, Cham, pp 213–223
Reckwitz A (2002) Toward a theory of social practices: a development in culturalist theorizing. Eur J Soc Theory 5(2):243–263. https://doi.org/10.1177/13684310222225432
Savoia A (2011) Pretotype it. Make sure you are building the right it, before you build it right. https://www.pretotyping.org/uploads/1/4/0/9/14099067/pretotype_it_2nd_pretotype_edition-2.pdf. Accessed 25 June 2020
Simon HA (1969) The sciences of the artificial. MIT Press, Cambridge
Vistisen P (2016) Animation-based sketching. An explorative study of how animation-based sketching can support the concept design of non-idiomatic digital technologies. PhD thesis dissertation, Aalborg University, Aalborg

Chapter 5
Designing for the Level of 'Service as Infrastructure'

Abstract This level suggests a view of services as an aggregation of human, organisational and technical factors to generate potential value. The activity of designers at this level has been widely studied; therefore, this chapter focuses on the specific design capabilities that come into play when working at this level.

5.1 What Are Designers Designing at This Level?

Designing the infrastructure for value co-creation means creating the conditions for customers to create value. This involves designing physical or virtual spaces in which the interaction will happen, orchestrating the timing for coordinating the contribution of various stakeholders, and visualising opportunities for customers to aggregate the available resources to create meaningful value.

In fact, the designer's activity at this logical level is much more concrete than at the value-co-creation level. Shostack (1982) used the analogy of a battery to describe the value proposition created at this level. Designing services at this level, according to Shostack, can be compared to the power stored in a battery. The battery has potential power, which can only be used once the contact is activated. Likewise, the service designed at this level consists of material and immaterial components aggregated in a way that creates potential value. Such potential can only become real value with the interaction with the customer.

A bank is an organised structure that is designed and made available for customers, but only when customers enter the bank (or when they access its online services) does the value of the service become evident.

The aggregation of material and immaterial components is based on social, organisational or cultural instances that designers need to interpret and represent in the service proposition. The combination of such heterogeneous elements—what Casper and Latour (2000) would define as *socio-material assembly*—could be offered by a service provider or a constellation of actors, and the possible configurations are the

N. Morelli et al., *Service Design Capabilities*, Springer Series in Design and Innovation 10, https://doi.org/10.1007/978-3-030-56282-3_5

outcome of an activity of design, which puts together the elements on the basis of knowledge about customers, technical issues, organisational instances and cultural contexts.

Designing a bank service means arranging people, competences, technologies, and organisational instances that will support the interaction between the customer and the bank, both in the front office (i.e. in the very moment in which the interaction happens) and in the back office (i.e. in each action generated in the organisation of the bank, as a prerequisite or a consequence of the interaction with the customer).

At this level, service design consists of generating the physical or logical context, which represents the ecosystem in which value is co-created. This means designing service organisations, public administration functions, service platforms or even policy instruments that all deliver specific service propositions. This implies a collaboration between designers and private or public institutions, such as firms, commercial platforms, healthcare or municipal organisations, or taxation offices.

5.2 The Role of Design at This Level

The creation of services as the infrastructure for value creation requires the use of expert knowledge and capabilities concerning technical issues (e.g. knowledge about specific software), system organisation (e.g. understanding the structure of a company or an institution in respect to its mission), other specialised knowledge (e.g. about logistics, healthcare, et cetera), and social and behavioural knowledge concerning the customers of the service, their preferences, their attitudes and needs.

None of these capabilities are in fact specific to the designer, but rather they concern diverse disciplines, from IT to marketing. The role of design, however, consists of orchestrating such knowledge, possibly bringing together perspectives that may not be represented in the mix of expert capacities in a design team. Designers, for instance, often bring to the table user-related perspectives, as their expertise and education often include methods and tools for understanding customers. In addition, designers' user-related perspectives may be the result of their own involvement in co-creation processes.

Designers facilitate the process of creating such infrastructure through their capability to represent logical architectures, interactions, time sequences and experiential elements of the service. The contribution of designers may consist in generating *tangible* elements for negotiation among stakeholders that interact in different moments, bringing about different cultures and practices into the system. By tangible elements, we mean elements that can be materially perceived or logically understood by the stakeholders. This highlights the role of orchestrator that designers can have in the development of new services.

5.3 Design Capabilities at This Level

Because of their role as *orchestrators*, designers' capabilities must include *connective* knowledge between different disciplines. Designers do not need to be coders, but they do need to know what to do with the results of the coding. They do not need to be managers, but they also need to know what the managers need for articulating their strategies or how to translate managers' idea of a service into concrete instructions that will make the service possible. In addition, they do not need to be marketing experts, but they do need to know how to elicit information about the customers' needs and expectations about the network of stakeholders. This will contribute to value creation and the existing value and motivations that keep the service system together.

The expert capabilities needed in the perspective of design for infrastructuring include:

- *Open problem solving*, which is an approach to problem solving that involves the creation and evaluation of multiple alternatives. This would support the organisation of service platforms, business models, or the interaction between customers and the technological components involved in the service architecture.
- *Building logical architecture*, which refers to the ability to create value by relating each individual element to the architecture of the service, thus generating structures and frameworks that link and combine different knowledge into service propositions.
- *Vision building*, which is the ability to model and visualise solutions into coherent representations of possible futures. Such visualisations include the representation of interactions among the actors, systemic maps and representations of business opportunities.
- *Addressing the context*, or more specifically, the ability to align solutions to their social/cultural context or to existing policies or corporate missions.

5.3.1 Open Problem Solving

If we accept the axiom that the value in service design is only created by the beneficiary (Vargo and Lusch 2004, 2008), we must understand that the design of infrastructure that supports value co-creation must remain open to a range of possible alternatives: the different ways that beneficiaries or other actors may interpret the service, the different cultural backgrounds of the beneficiaries, and the different conditions of the interaction. We must also take into account that, as mentioned, value in services is *potential* value—it is only released when interacting with customers. The action of expert designers can be more or less prescriptive in respect to the behaviour of different actors in the service ecosystem: it would give precise indications of time and the sequence of actions for services based on mechanical or safety-related procedures—for example, the procedures for accessing or undergoing medical treatments—or provide a framework of behavioural indications, or possible interactions, much like the interaction rules in service platforms.

5.3.2 *Building Logical Architecture*

Infrastructuring means aggregating a number of heterogeneous elements, including objects, technologies, knowledge, people and spaces, according to a logical organisation that can efficiently and satisfactory address the needs of users, communities and contexts. This involves understanding the relations between heterogeneous elements, such as the interaction between people and technologies, power relations, social relations, and knowledge exchange. Every new service consists of an *ecosystem*, which implies a systemic approach to the organisation of such interactions. When building the logical architecture of a service, an expert designer should be able to navigate among different kinds of knowledge and provide opportunities for this knowledge to combine in many possible ways.

5.3.3 *Vision Building*

The organisation of infrastructure implies the capability of figuring out its possible configurations and the way such configurations will be interpreted or used by the service beneficiaries or other actors in the service ecosystem. Vision building is also vital in negotiating possible configurations of the service with service providers and other stakeholders participating in the service because it is a way to propose concrete images or experiences of the service proposition, including business, organisational or emotional aspects, before any element of the service is in place. This capability gives designers a role as facilitator in the process of creating the service proposition, as such visions support the negotiations and alignment of multiple views and interests.

5.3.4 *Addressing the Context*

Service propositions are always related to specific contexts. While the organisational structure of a service can be replicated in different contexts, it is essential that any configuration of the service provides a valid service proposition in each specific context. This means the capability to understand how the interaction among the stakeholders of a service can be adapted to specific situations or contextual conditions.

Table 5.1 Designing as infrastructuring: the capabilities needed and most common tools used

Objects of the design activity	Characteristics of the design activity	Examples	Capabilities required
Services Service platforms Policy instruments	Organisation of services Support to public administration	Healthcare services Fast food chains Franchising	Building logical architecture Vision building Addressing the context
		Cooperative platforms (social housing) Commercial platforms (AirBnB, Uber, Facebook, etc.)	Building logical architecture Open problem solving

Designers can work on supports, such as cards, templates, guidelines or procedures, that help the actors in each specific context to generate relevant value out of a service proposition. For example, cards might be used to help the communication between different stakeholders, whereas journeys can be used to support users in figuring out how to interpret and adapt a service structure. Such elements must be part of the structure of the service and complement the functional or organisational components of the service ecosystem.

5.3.5 Summary

Table 5.1 summarises the expert design skills needed in the perspective of designing as infrastructuring, including some of the most common tools used.

5.4 Cases

5.4.1 Platform-Based Services: Social Housing

Introduction The pressure of the housing emergency in many countries and the increasing price of real estate on the housing market in bigger cities require solutions to provide an attractive place to live to the more economically disadvantaged part of the population that would otherwise be priced out of certain urban areas.

Public housing policies have often proved insufficient in satisfying the housing needs of the economically weakest part of the population, and in some cases, large public housing developments have become ghettos, where the social and health conditions of the inhabitants are substandard, thus increasing social inequality in cities. In some countries, non-profit organizations have joined the effort to provide affordable and decent housing for all, thus generating a form of social housing that considered not only the basic need for housing but also the social need to create an attractive, fair and social-minded environment within the housing estates.

In addition to the financial effort for the construction or adaptation of the buildings, social housing requires means testing to verify the access rights to social housing, but in more recent history, some social housing organisations have also created platforms to accompany citizens in a process of integration and co-creation of the social environment in the housing estate. Those platforms aim at accommodating the functional needs in the shared housing estate environment with the habits and social needs of its inhabitants.

During a number of experiences with social housing, the *Fondazione Housing Sociale* (FHS—Social Housing Foundation) in Milan developed a platform to guide new social housing settlements or else ensure that existing social housing settlements maintain a healthy and positive environment.

The platform includes a number of tools (e.g. social cards) to support the settlement of the inhabitants and negotiate a way to live together, by deciding how to use common spaces, getting to know each other. The *stacking* plan (Fig. 5.1) allows inhabitants to indicate their family composition, social capabilities, age and preferences to book common spaces and facilities and organise events. The platform also includes a roadmap for the creation of new communities that includes pre-settlement periods—in which it is possible to figure out what kind of life is possible to live in the housing—and the actions to regulate everyday routine.

Definition of the service nature The social housing platform is an innovative version of a new family of platform-based services, in which the value proposed consists of a number of supports for negotiation, while the definition of the nature and the form of the value to propose is left to the platform user.

Platform-based services are becoming extensively diffused, and they target various aspects of our life, such as tourism accommodation, car sharing, tool sharing, mutual help, and food preparation and consumption.

The platform service provider mediates with different actors, leaving different ranges of options for the platform users to control the service and decide on the value to be created in the platform (Choudary 2015). The platform model takes different forms, ranging from neo-capitalistic configurations, in which a company has the power to control the flows of information and economic transactions, to distributed models, such as the cooperative platform of this case, in which decisional, power and economic transactions are controlled by a community of actors (Scholz 2014).

Role and challenge for designers The role of designers in this specific project was to ideate and create the tools to support the negotiation process between the different actors. Designers directly participated in the meetings, proposing tools such as cards, templates, roadmaps and other community-building procedures. However, the direct participation of designers in the negotiation process is not always the case in platform-based services. Direct participation makes it possible for designers to understand the dynamic of negotiation and the way different practices and routines are proposed, compared and framed, according to the specific conditions of the

Fig. 5.1 A stacking plan is a synthetic representation of the families that are going to live in the same social housing building, including the composition of the families, their preferences and their contact information. Reproduced from Ferri (2016)

housing and the neighbourhood. In other cases, direct interaction is not possible, and different strategies are proposed instead.

The experience acquired by direct participation is often 'codified' in the tools proposed (cards, templates, roadmaps) in a way that can be independently used by new communities, without the presence of designers. In other cases (e.g. in worldwide platforms for house rentals), the negotiation tools are progressively developed, on the basis of the feedback from the platform users, the problems and complaints emerging in the platform and most probably the direct experience of the service providers. In this case, the designer's effort is still to codify and 'pack' various suggestions, providing inspiration or support and creating consistent stories, checklists, or practical indications for the users.

Design capabilities involved The work of platform-based services is based on the creation of practical support, from apps, cards, canvasses and templates to roadmaps. The capabilities involved are:

- *Open problem solving*: designers do not solve the problems of the social housing community but rather generate a framework in which problems can be better understood, framed or articulated in simpler parts. It is up to the community to employ the designer's support to solve specific problems.
- *Building logical architecture*: in platform-based services, this capability consists of identifying key roles or matching different practices and competences to be framed in an organised structure. In this case, the capability of architecture building uses a real-life example of architecture, the social housing building, to study and define the way different families will live together.
- *Vision building*: in this context, this capability consists of proposing leading inspirations about possible directions that the community of social housing inhabitants can explore. This is particularly evident in large platform services where the participation to the platform depends on the service provider's and the designer's ability to give their customers a good view of the work they have to do together and the results they can gain from participating.
- *Addressing the context*: platform-based services providers often propose an 'unfinished' solution that is meant to adapt to different contexts. However, it is important that the design of the platform can capture and incorporate the characteristics of the context within the mechanisms of exchange among the platform users. Cards (as in Fig. 4.1) or roadmaps to consolidate the community of inhabitants of each social housing dwelling are the elements that, in this case, supported the formation of different communities in different contexts.

5.4.2 Hackathons in the Open4Citizens Project

Introduction The hackathon format has emerged in the past few years as a successful format to gather participants and jointly work on issues of common interest. At the end of the 1990s, hackathons were niche events mostly organised and attended by open-source software developers (Briscoe and Mulligan 2014). However, today the format of the hackathon—appropriated or reinvented by design, innovation and start-up communities—is increasingly used to organise events attended by a variety of participants (including non-expert programmers) and aimed at different scopes, from exploring new production processes (Tanenbaum et al. 2014) and tackling social issues through humanitarian technology (Linnell et al. 2014) to prototyping a new generation of services and new ways of commoning (Morelli et al. 2017a, b).

The latest case is that of the EU-funded Open4Citizens project, which, through new forms of collaboration between public authorities, citizens, interest groups, local businesses and IT experts, aims to (1) aggregate communities around open data, (2) develop a set of practices and infrastructures for using such data and (3) generate new public and private services.

Open4Citizens was articulated in several pilots that worked on various challenges ranging from migration to tourism within a shared framework of design processes, tools and methods to increase citizens' awareness of open data and to engage them in the creation of new solutions for their everyday problems.

The citizens' participation in the pilots was mainly supported through the organization of hackathons (Morelli et al. 2017a, b), which would bring together a local community with shared interests in specific problematic areas, such as public roadworks in urban contexts or issues related to healthcare, integration, pollen pollution or the regeneration of parks. Within the Open4Citizens project, it was soon realised that in order to mobilise the relevant stakeholders for a successful hackathon event, a 'hackathon campaign' needed to be launched, which started a few months before the actual event. Many of the activities of the designers were then devoted to building a solid interaction that would give relevance and effectiveness to the hackathon. The design team's work aimed at (1) aggregating a number of relevant stakeholders around the event, (2) raising critical issues with the relevant stakeholders (is the data ready? Is the challenge relevant?), and (3) promoting a new generation of public services based on open data.

Definition of the service nature In the Open4Citizens hackathons, a big effort was spent managing and keeping the stakeholders involved and securing their contribution to the hackathon. In this way, the designers were 'infrastructuring', given that they were supporting the creation of a new ecosystem around the open data commons.

Role and challenge for designers The role of the designer in this project has mainly been to ideate and create tools to support the collaboration among the hackathon participants (the hackathon toolkit), but most importantly, it is to ensure the support of the relevant local stakeholders, from public authorities to the data owners. While the tools that supported the event could clearly and easily be applied in the various contexts, the mobilisation of stakeholders and communities took very different paths in the different contexts.

Design capabilities involved To mobilise the needed stakeholders, several tools have been used as practical support, such as canvasses, mind maps, stakeholder maps, cards, et cetera. Of all the capabilities used, some were more relevant than others:

- *Open problem solving*: this activity characterises the very nature of hackathons in which problems emerge and are clearly defined through the interaction of the different stakeholders. The hackathon was used as a physical and logical place for conversation and negotiation rather than to find a finite solution.

- *Vision building*: the designers had to convince and involve public authorities and other private/public organisations in order to share a common vision about the use of Open Data. The pre-hack work was often dedicated to creating such visions, while the hackathon itself was an opportunity to build shared visions, where the outcome of a hackathon is often regarded as a possible representation of such vision.
- *Addressing the context* is crucial to understand what kind of value proposition the different stakeholders can bring to the table or can contribute to create. This capability has been crucial in the pre-hack activities, to define the focus of the co-creation exercise in the hackathon and to aggregate the most relevant actors that could interpret the instances from the context.

References

Briscoe G, Mulligan C (2014) Digital innovation: the hackathon phenomenon. Working paper no. 6. Creativeworks, London. https://qmro.qmul.ac.uk/xmlui/handle/123456789/11418. Accessed 25 June 2020

Casper MJ, Latour B (2000) Pandora's hope: essays on the reality of science studies. Contemp Sociol 29(5):754–756. https://doi.org/10.2307/2655272

Choudary SP (2015) Platform scale: how an emerging business model helps startups build large empires with minimum investment. Platform Thinking Labs

Ferri G (ed) (2016) Starting up communities. Un Design-Kit per l'Abitare Collaborativo. Mondadori, Milan

Linnell N, Figueira S, Chintala N, Falzarano L, Ciancio V (2014) Hack for the homeless: a humanitarian technology hackathon. In: IEEE global humanitarian technology conference (GHTC 2014), San Jose, 10–13 Oct 2014. https://doi.org/10.1109/GHTC.2014.6970341

Morelli N, Aguilar M, Concilio G, de Götzen A, Mulder I, Pedersen J, Torntoft LK (2017a) Framing design to support social innovation. Des J 20(Sup 1):171–184

Morelli N, Mulder I, Concilio G, Pedersen JS, Jaskiewicz T, de Götzen A, Arguillar M (2017b) Open data as a new commons: empowering citizens to make meaningful use of a new resource. In: Kompatsiaris I, Cave J, Satsiou A, Carle G, Passani A, Kontopoulos E, Diplaris S, McMillan D (eds) Lecture notes in computer science, vol 10673. Springer, pp 212–221

Scholz T (2014) Platform cooperativism vs. the sharing economy. https://medium.com/@trebors/platform-cooperativism-vs-the-sharing-economy-2ea737f1b5ad#.oztuau1k4. Accessed 25 June 2020

Shostack LG (1982) How to design a service. Eur J Mark 16(1):49–63

Tanenbaum K, Glen Tanenbaum J, Williams AM, Ratto M, Resch G, Gamba Bari A (2014) Critical making hackathon: situated hacking, surveillance and big data proposal. In: CHI'14 extended abstracts on human factors in computing systems, p 17

Vargo SL, Lusch RF (2004) Evolving to a new dominant logic for marketing. J Mark 68(1):1–17

Vargo SL, Lusch RF (2008) Service-dominant logic: continuing the evolution. J Acad Mark Sci 36:1–10

Chapter 6
Designing for the Level of 'Service as Systemic Institutions'

Abstract Every action is framed in a broader landscape dominated by a system of values, rules, and cultural, social and political premises. In this landscape, change is slow, evolutionary, and lies outside the control of a designer. Nevertheless, designers can play a role in triggering change and possibly steer it in preferred directions. This chapter illustrates the design capabilities that are most relevant for design action at this level.

Every action that produces change occurs in an institutional landscape. By 'institutional landscape', we mean the system of values, rules, and social, cultural, economic and political premises that frame the change and facilitate the changes that are consistent with this framework, while hindering those that are not (Koskela-Huotari and Vargo 2016; Vargo and Lusch 2015).

It is easy to imagine that, like natural landscapes, the institutional landscape may seem static or changing very slowly. Change and innovation at this level happens by slow and evolutionary movements influenced by several factors, including, but not limited to, human action. For instance, the institutional landscape is influenced by changes in institutional settings, such as cultural, scientific and technical knowledge, politics and social trends. Human action, and consequently design action, is therefore unlikely to produce direct and controllable institutional changes. Nevertheless, human action can still influence such changes, as shown by several examples of purposeful social constructions aimed at influencing this level. This is the case, for example, of healthcare reforms such as the welfare state reform in UK[1] or Obamacare in United States,[2] the Australia tax reform (Terrey 2012), the construction of the American electricity system (Bijker 1995), or the psychiatric reform in Italy (Manzini 2015). It is worth noticing that, besides the Australian Tax Reform,

[1]The welfare state reform inspired the Beveridge Report (1942) and was initiated by the first post-WWII government inspired by the work of John Maynard Keynes and William Beveridge.

[2]The Patient Protection and Affordable Care Act (PPACA), also known as Obamacare, was approved in the United States in 2010.

N. Morelli et al., *Service Design Capabilities*, Springer Series in Design and Innovation 10, https://doi.org/10.1007/978-3-030-56282-3_6

the main actors in these cases were not expert designers,[3] even though their action could be defined as a design action, in so far as it was aimed at an intentional change. Therefore, it is important to understand what role designers and design action can have in the transformation of the institutional landscape.

6.1 What Are Designers Designing at This Level?

Working on changes in institutional landscapes implies a very broad scale of intervention. This approach is quite new for the design discipline, and the discussion about the role and contribution of designers and design—seen as a discipline or a body of knowledge—is still quite open (Koskela-Huotari and Vargo 2016; Vargo and Lusch 2015; Vink et al. 2017).

In his classification of the areas for design thinking, Buchanan (1992) listed the '*design of complex systems or environments for living, working, playing and learning*' as the widest area of design thinking. At this level, design activity faces a high level of complexity and is often geared towards solving *wicked problems* (Buchanan 1992; Rittel and Webber 1973), interacting with multiple disciplines, and dealing with multiple dimensions (cultural, technical, social, environmental, etc.). For this reason, Banerjee (2014) proposes that a fifth order could be added to Buchanan's classification: the design of *large-scale transformations*. When focusing on this kind of transformation, he indicates three types of design intervention:

- Designing new systems and scaling-up paradigms for unfamiliar and scaled problems, for example, by reducing the energy footprint of an entire nation by a significant percentage through a combination of approaches.
- Transforming the behaviour, roles and the relationships of the constituent stakeholders within the ecosystem, including non-human elements such as resource flows and natural systems, for example, an intervention to solve the food security problem through the simultaneous engagement of banks, government, agro-industry, farmers, small business enterprise, telecom companies and non-profit companies.
- Transforming the behaviour, outcomes and trajectories of the larger ecosystem, for example, by changing the way in which our institutions, civic societies, industry and government work so as to get entirely new trajectories regarding social, economic and environmental issues, while, at the same time, building a more resilient, more shock-resistant future. This could be achieved, for example, by combining healthcare, social, economic and environmental policies, to generate

[3] The Australia Tax reform was the result of an explicit intention to integrate design and designers into the transformation of the taxation system. The construction of the American electricity system was instigated by the action of Thomas Alva Edison and his company. The psychiatric reform in Italy was instead the result of the policy action of a psychiatrist, Franco Basaglia, who was the major inspirator of the legislative reform.

a system that is more resilient to pandemic crises, while re-localising production, and reducing the environmental impact of mobility flows.

In operative terms, designers contribute to generate policies, as in, sets of principles for action that inform the government of a large community of individuals. When working at this level, design can be essential to trigger or visualise possible change, or to highlight key elements that can play a role in large-scale transformation. It is important to know, however, that neither designers nor any other actors involved in such transformation can ever have full control over the change they are sparking.

The contribution of designers mainly consists of exploring possible futures, generating visions, creating scenarios and devising strategies for scaling-up promising initiatives; but at the same time, design activity is needed to align changes in infrastructure or even everyday value creation systems to the changes in the institutional system.

6.2 The Role of Design at This Level

According to Banerjee (2014), working for such transformation means, from the design perspective, considering two possible (and not alternative) strategies. The first is vertical co-creation, which consists of aligning and amplifying any promising change at any level to highlight an emerging direction for transformation at the systemic level.

A strategy of this kind is related to the question of the scalability of small initiatives, from local and limited contexts to wider contexts. Manzini and Rizzo (2011) propose that this can be done by defining framework programmes within which small experiments can be allowed. If particularly relevant for the context, such experiments can be amplified to larger scales. A different approach is proposed by Morelli (2015), who suggests identifying and codifying the structure of the socio-technical ecosystem of successful initiatives on a small scale. Such a structure—not the ecosystem per se—could then be reproduced in different or wider contexts.

The second strategy, horizontal co-creation, consists of creating shared perspectives and working across different disciplinary boundaries in order to generate holistic and shared views of possible transformation. These views have the potential to inform policies and policy instruments and will influence the development of new services or inform new cultures and individual behaviours.

A design approach applied to large-scale scenarios can contribute to defining the vision of what direction is desirable and feasible for the development of large-scale issues, thus suggesting mission-oriented innovation policies. This process would generate systemic public policies that draw on frontier knowledge to attain specific goals (Mazzucato 2017, p. 8). At the same time, the design contribution could help create operative perspectives for translating such visions in concrete design actions. This approach could be useful in approaching 'grand social challenges', such as the redefinition of social and economic ecosystems after a pandemic, the conversion towards a more sustainable society, inequality and youth unemployment, and to provide strategic direction for funding or supporting policy instruments, organisational arrangements, public services and cultural transformations.

6.3 Design Capabilities at This Level

The strategies outlined above imply different capabilities: vertical co-creation implies the capability to see or build a consistent framework to describe future perspectives that are instrumental when defining preferred future directions of development. Looking towards the future does not necessarily mean projecting the present onto a future direction (*forecasting*), but rather it can imply the opposite process of imagining and visualising preferable futures and projecting them onto the present—to highlight preferred trajectories and changes that will lead towards that future. This activity is defined as *backcasting* (Holmberg 1998). Backcasting has been used by companies or public authorities to figure out long-term strategies or policies and, on the basis of this, orient present and medium-term action.

In relation to scalability, vertical co-creation implies the capability to recognise and map complex ecosystems (e.g. the actors of a healthcare system or a community) in order to amplify its structure or reproduce the ecosystem in different contextual conditions (Morelli 2015; Manzini and Rizzo 2011).

Horizontal co-creation implies a design capability to generate shared visions and negotiate the transformation among a number of stakeholders that have different motivations, perspectives, interests and cultural frameworks. Design capabilities are also relevant to align such shared vision to the present reality, thus suggesting operational ways of addressing them.

Therefore, certain design capabilities are needed for designers to contribute to large institutional changes:

- *Vision building*: at this level, this implies the capability to aggregate different components of a complex reality into consistent, credible and desirable scenarios.
- *Modelling*: this refers to the credible and consistent simulation of what future institutional configuration will look like or imply in operational terms.

- *Working at different levels of abstraction*: this means being able to relate large-scale changes, such as policy changes or new regulations, to the infrastructure that will support such changes and figure out how those transformations will impact the everyday value-creation processes.
- *Addressing the context*: this refers to the ability to look at the broader picture, i.e. the technical, economic, social and environmental conditions in which the actors interact.

6.3.1 Vision Building

This capability refers to the use of scenarios or models that describe possible futures and provide a framework to evaluate the change and the various criteria that highlight the main factors of change. To work on large institutional changes, a designer has to be able to perceive and elaborate the signals and the needs of large cultural changes (e.g. the attitudes or initiatives towards social innovation, broad transformation based on technological advancements, the need for a sustainable development) and aggregate them into relevant representations of possible, feasible and desirable transformations. Visions are also essential to support the aggregation of actions otherwise dispersed in different domains thus creating a consistent image of a possible future transformation. For example, visions that aggregate changes in the healthcare system, in the welfare system and in the localization of production may address the transformation towards a system that is more resilient to a economic or pandemic crisis.

6.3.2 Modelling

When working on large institutional changes, modelling is intended as the capability to generate consistent and tangible descriptions of possible futures. These possible futures are visualised through operative aspects of the proposed change, even before all the strategic or operational aspects have been clearly defined. The resulting visualisations serve as an effective tool for policy makers to set up their policy objectives, or for large communities for recognising possible development paths and working in that direction.

6.3.3 Working Across Different Logical Levels

To be effective or to generate an actual impact, large-scale transformation should link to transformations in the smallest scale of the social or technical ecosystem, such as transformation in services, in infrastructures, and even in everyday interactions and value co-creation processes. When large-scale transformation is related to the activity of policy making, this capability is referred to as policy delivery. In the opposite direction, this capability refers to pointing out small-scale transformations that have great potential for developing and working towards scaling them up[4] in order to contribute to larger institutional changes. The designers' work can therefore be crucial for aligning the different scales of innovation possibly working on scalability, or in the opposite direction, on policy delivery.

6.3.4 Addressing the Context

At this level, understanding and addressing the context for innovation or design action means getting a wide perspective of complex systemic issues, considering a number of actors, critical factors, and technical, economic, social and environmental conditions. Defining future directions for institutional change requires a clear view of the present, which is not always easy. Institutional systems have a degree of complexity that does not allow for an unequivocal interpretation or a clear definition of the problems to address; nevertheless, the need for action requires that designers are able to map, interpret and highlight critical issues that could be taken as the *handle* to *grab* and modify such a complex reality. Critical logical maps in addition to narrative techniques are often used in those cases to highlight relevant issues and compare different phenomena and trends.

6.3.5 Summary

Table 6.1 summarises the design capabilities that are needed to work on systems and institutional structures.

6.4 Cases

6.4.1 The Reform of the Australian Taxation System

Introduction Taxation is at the basis of the financial system of a country, and it is a pervasive component of business and private life. Depending on the country, taxation systems are more or less transparent in indicating the way taxes contribute to the economic management of a country and more or less complex and easy to understand for citizens.

---

[4]With the term, 'scaling up', we indicate the vertical process of institutionalisation of innovation initiatives, as opposed to the horizontal process of 'scaling out', which refers to the expansion of innovation initiatives or the enlargement of their base of adopters (Hancock 2003).

Table 6.1 Service design capabilities for working on systemic institutional changes

Objects of design activities	Characteristics of design activity	Examples	Design capabilities needed
Policy documents (intent documents, white papers, quality assurance documents, design briefs)	Suggesting directions of change in institutions, government organisations	Long–medium term Sustainability plans Healthcare reforms Tax reforms (Terrey 2012) Urban health (Geels 2010) Edison's electricity system (Bijker 1995) Psychiatric reform (Manzini 2015) Gas stove systems (Cowan 1987)	Modelling Working across different logical levels Addressing the context
Scenarios, models	Triggering change in local communities, and institutions	DoTT 07 (Manzini and Rizzo 2011)	Vision building
Public innovation spaces	Triggering change in institutions, local communities, governments	Policy labs Living labs	Working across different logical levels Vision building

The transparency of the relation between the tax we pay in our country and the public goods and services we can access, such as education, transport, health care, policing, is key in democratic systems, as it is part of the legitimation of public authorities. At the same time, the complexity of contemporary societies requires a more and more articulated taxation system, that makes it harder to understand and manage for the majority of citizens.

A major change in the Australian taxation system, the introduction of the goods and service tax, was the trigger for a substantial revision of the way citizens experienced the system. For this reason, the Australian Taxation Office (ATO) decided to use design to make the government financial policy clearer, to translate strategic design of the office into action, and to make the tax system easier to access, cheaper, and more personalised for citizens.

The office proposed a vision aimed at improving its interaction with taxpayers. The ATO organised a programme involving the progressive assimilation of design thinking into the strategic procedure of its management. The programme originated from three conferences that introduced two big intellectual challenges for the organisation: one consisted of understanding the design approach, and the other involved building a design capacity within the organisation, which would give a better capability to navigate among the general principles and strategy, the different projects, and the tools to be used in each project.

Definition of the service nature When working on the new tax system, the ATO task force intended to encourage compliance with the taxation system by building a more efficient communication and management strategy, based on a tighter collaboration between taxpayers, intermediaries and the system and by offering a number of products and services to assist business operators (Commonwealth of Australia 2003).

The introduction of design in the ATO implies a higher focus on the taxpayer, making sure, however, that such a focus is consistent with the general principles of the organisation. Over the financial year, the taxpayer has to put together a number of direct or indirect interactions with the tax system: keeping tax-related receipts, receiving instructions, filling in forms, receiving payment summaries, filing their tax return, and finally, making a payment. Highlighting such operations is already moving the focus away from the functionalities of each of those functions to that of the whole experience (Body 2008, p. 58).

The transformation within ATO consisted in creating a design capability to assist teams in all areas. Such a capability could be built as an internal resource that would ensure that design action refers to certain general principles, including user centricity, visibility of the processes, collaboration, shared understanding, consistency, user centricity and balance between tax system integrity and user experience.

Role and challenge for designers The ATO is the administrative component of the whole Australian Taxation System. The function of design in this project was therefore not meant to change the whole taxation system but rather to facilitate change in a clearly defined institutional organisation.

The need for ATO to include design as a new internal capability emerged after a number of negative evaluations of the tax reform in Australia (Terrey 2012, p. 4). As a result, the management of the ATO decided to engage Prof Richard Buchanan from Carnegie Mellon University as a mentor to spark the process of the progressive assimilation of design capabilities in the organisation. The challenge for the organisation was not only to understand the benefit of a design approach but also to introduce a new day-by-day practice of design involving a new set of tools, methods and strategies guided by a specific, design-led methodological approach. The project team decided not to push design top-down to all levels of the organisation, and instead, offer design as a service provided by a design team working on the parts of the organisation where the need for a design approach was more evident.

Design capabilities involved in the project Design capabilities were used for facilitating change within the taxation office's practice by synthesising solutions and for communicating in various stages of the process, from user requirement visualisation to policy implementation. In this case, three design capabilities have emerged as most relevant:

- *Addressing the context*: the starting point of the redesign process was the first conference, where a design perspective was proposed, which moved the focus of management action from the organisation to the user. This was done by generating a narrative description of the taxpayer's pathway through the system, which gave the possibility to describe the details of each interaction between the taxpayers and the taxation office.

- *Working across different logical levels*: Starting from the analysis of the taxpayer's experience, the ATO needed to link this knowledge with new policies. This is possible through a collaboration between ATO and the treasury since the early stage of policy formulation, in order to link design implementation and new legislation and to meet community concerns (Fig. 6.1).
- *Vision building*: this capability is used to produce documents of intents, which describe the intended process, with the aim of generating a common understanding within the team. Visualisation was also used to represent the design process, to describe the way different projects could run at the same time and to describe the integrated tax design process, which highlighted the interaction between the ATO and other institutional actors, such as the treasury and parliamentary offices (Fig. 6.1).
- *Modelling*: the activity of prototyping was used to identify users' requirements. During the stages of the process that followed, a simulation centre was built in Brisbane, which made it possible to prototype the interaction between the taxpayer and the system.

6.4.2 Living Labs

Introduction and definition of the service nature According to Mark De Colvenaer, 'Living Labs are an open innovation ecosystem where partners or stakeholders from different backgrounds can work together to find solutions to a defined challenge' (cited in Hellström Reimer et al. 2012, p. 23).

Generally, Living Labs work as environments geared towards carrying out and supporting collaborative projects with external stakeholders. In some cases, Living Labs have a tangible, physical dimension. Over the past ten years, the Living Lab, Fabriken, located in Malmö in the south of Sweden, has been in operation, offering the public free access to a workshop equipped with a laser cutter, 3D printer and other prototyping tools (Seravalli and Simeone 2016). In addition, Fabriken also hosts a variety of events open to the public and organised either by the Living Lab managers or by the local community of users. Examples of such events range from micro-crowdfunding dinners all the way up to workshops on inclusive and sustainable fashion, drop-in art sessions and seminars on self-publishing.[5]

In some other cases, Living Labs do not have a physical space but rather operate as an organisational framework that brings together stakeholders and engages them in co-creation activities, typically through a series of events such as hackathons, workshops and seminars. This is the case, for example, of the 6 Living Labs set up by the EU-funded *iScape* project.[6] Over a period of three years, these labs connected

---

[5]https://stpln.org/calendar. Accessed 3 April 2020.

[6]https://www.iscapeproject.eu/. Accessed 3 April 2020.

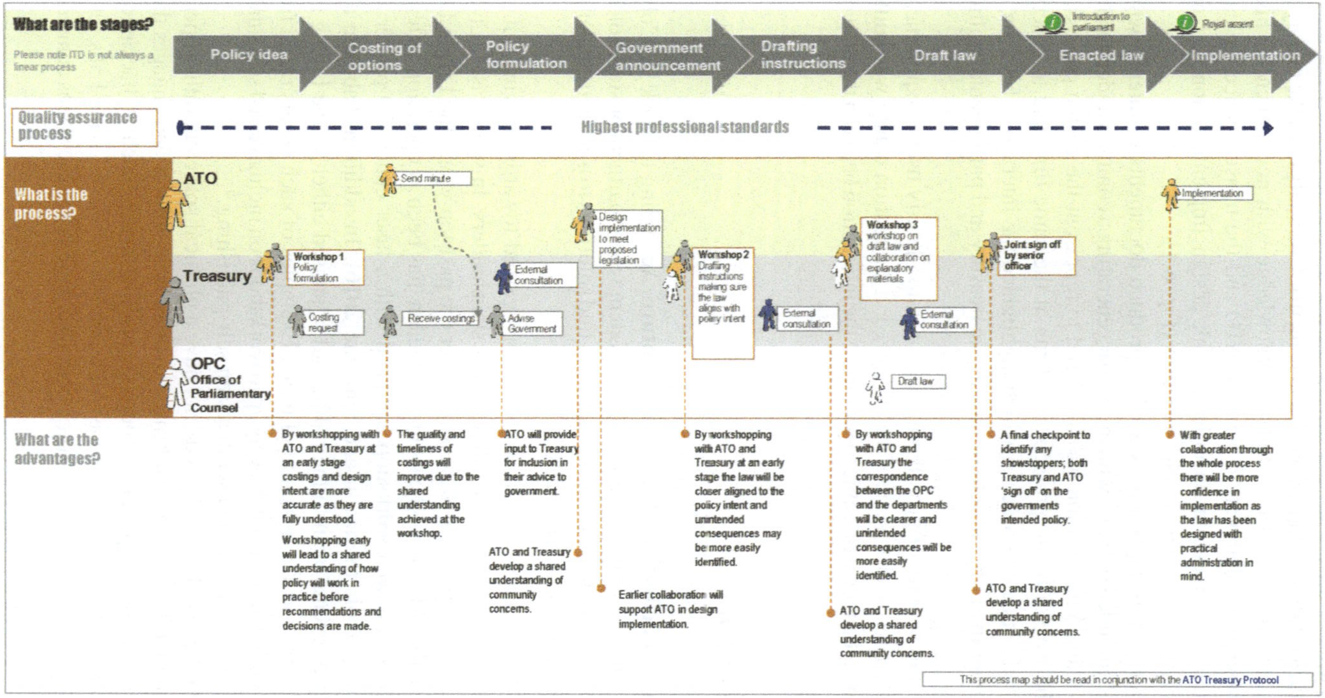

Fig. 6.1 The process map visualises the way policies are implemented, specifying the stages and the interactions to align actions and policy intents. *Source* Australian Government (2012, Appendix A)

a great variety of stakeholders, 'facilitating collaboration and sharing of multidisciplinary knowledge and experience to advance air pollution remediation strategies and solutions.'[7]

Role and challenges for designers The support that designers can offer to Living Labs can be articulated into two main areas: (a) the designers can contribute to the initial set-up of these labs, for example, imagining what kind of physical space and equipment are needed (if any), their governing and organising rules and the models to engage external stakeholders; (b) the designers can contribute to the daily operations of the living lab, for example, organising specific design-based activities (such as hackathons or prototyping sessions) to actively involve external stakeholders. However, the participation of designers in the creation of living labs also concerns the definition of the strategic role that the Living Lab can have in relation to how it can support local communities, institutions and urban ecosystems. Living Labs can be a Public Innovation Place, for example, places where citizens and public sector staff come and work together to create solutions which enable positive social change (Tassinari 2013).

Living Labs have become a quite popular format to set up co-creation environments. Since its formation in 2006, the European Network of Living Labs has labelled more than 440 Living Labs.[8] Some of them, like Fabriken, are particularly geared towards vertical co-creation. The activities of Fabriken are structured as a framework programme to upskill the local community and to support local initiatives and experiments that can be potentially amplified to larger scales. The case of the iScape Living Labs highlights a different strategy. Each lab was located in a different country and run by an organisation with specific and varying types of expertise (ranging from design and policy analysis for the built and natural environment up to transportation research, meteorological service, spatial planning, air quality and health). These Living Labs operated by both organising local events and collaborating at an international level with each other and with other stakeholders, including research centres, policy makers, companies and citizens interested in improving air quality. As such, the work of this network of Living Labs was mostly aimed at horizontal co-creation and at creating shared perspectives and European action plans.

The challenge designers face when involved in living labs concerns the complexity of such an initiative. Working with such a varied network of stakeholders, who may have different needs, interests and agendas, may speak different languages and may operate in different geographic and cultural contexts, can be quite challenging (Simeone 2016; Ehn et al. 2014). The design capabilities, however, can prove essential for facilitating alignment and vision building in such complexity.

Design capabilities involved Emerging from this case are two of the most relevant design capabilities:

---

[7]https://www.iscapeproject.eu/iscape-living-labs/. Accessed 3 April 2020.

[8]https://enoll.org/network/living-labs/. Accessed 3 April 2020.

- *Working across different logical levels*: given the different stakeholders involved in the various sites of operation, working through the format of the Living Labs means that designers should find or build broader frameworks to address, facilitate and orchestrate stakeholders' interaction. As such, rather than just focusing on one of the levels presented in this book (service as interaction, service as an infrastructure, service as a systemic institution), the designers instead work across various levels, mostly through operations of abstraction in which the problem or situation at hand is observed from different perspectives (or different levels of zooming). This facilitates the institutionalisation of bottom-up initiatives and, in the opposite direction, contributes to support policy delivery.
- *Vision building*: the capability to envision possible futures and to create consistent and structured visions of what that future could look like is one of the core components that brings stakeholders together and structures collaboration within a Living Lab environment and framework.

References

Australian Government ATO (2012) ATO treasury protocol. https://www.ato.gov.au/General/New-legislation/In-detail/ATO---Treasury-protocols/ATO---Treasury-protocol/?anchor=P58-8617#P58-8617. Accessed 28 June 2020

Banerjee B (2014) Innovating large-scale transformations. In: Bason C (ed) Design for policy. Routledge, New York, pp 71–86

Bijker WE (ed) (1995) Of bicycles, bakelites, and bulbs: toward a theory of sociotechnical change. MIT Press, Cambridge

Body J (2008) Design in the Australian Taxation Office. Des Issues 24(1):55–67. https://doi.org/10.1162/desi.2008.24.1.55

Buchanan R (1992) Wicked problems in design thinking. Des Issues 8(2):5–21

Commonwealth of Australia (2003) Australian Taxation Office, the cash economy under the new tax system. Department of Communications, Information Technology, and the Arts, Canberra

Cowan RS (1987) The consumption junction: a proposal for research strategies in the sociology of technology. In: Bijker WE, Hughes TP, Pinch TJ (eds) The social construction of technological systems: new directions in the sociology and history of technology. MIT Press, Cambridge, pp 261–280

Ehn P, Nilsson EM, Topgaard R (eds) (2014) Making futures: marginal notes on innovation, design, and democracy. MIT Press, Cambridge

Geels FW (2010) Ontologies, socio-technical transitions (to sustainability), and the multi-level perspective. Res Policy 39(4):495–510. https://doi.org/10.1016/j.respol.2010.01.022

Hancock J (2003) Scaling-up the impact of good practices in rural development. World Bank, New York

Hellström Reimer M, McCormick K, Nilsson E, Arsenault N (2012) Advancing sustainable urban transformation through living labs: looking to the Öresund region. In: Proceedings of the 3rd international conference on sustainability transitions, Copenhagen, 29 Aug 2012

Holmberg J (1998) Backcasting: a natural step in operationalising sustainable development. Greener Manag Int 23:30–51

Koskela-Huotari K, Vargo SL (2016) Institutions as resource context. J Serv Theory Pract 26(2):163–178

Manzini E (2015) Design, when everybody designs. MIT Press, Cambridge

Manzini E, Rizzo F (2011) Small projects/large changes: participatory design as an open participated process. CoDesign 7(3–4):199–215. https://doi.org/10.1080/15710882.2011.630472
Mazzucato M (2017) Mission-oriented innovation policy: challenges and opportunities. UCL Institute for innovation and public purpose working paper (2017-1)
Morelli N (2015) Challenges in designing and scaling up community services. Des J 18(2):269–290
Rittel H, Webber M (1973) Dilemmas in a general theory of planning. Policy Sci 4:155–169
Seravalli A, Simeone L (2016) Performing hackathons as a way of positioning boundary organizations. J Organ Change Manag 29(3):326–343
Simeone L (2016) Design moves. Translational processes and academic entrepreneurship in design labs. PhD dissertation, Malmö University. https://dspace.mah.se/handle/2043/21426
Tassinari V (2013) From welfare state to partner state: the case of welcome to Saint-Gilles. In: Manzini E, Staszowski E (eds) Public and collaborative, exploring the intersection of design, social innovation and public policy. Desis Network, pp 39–50
Terrey N (2012) Managing by design: a case study of the Australian Tax Office. PhD dissertation, University of Canberra
Vargo SL, Lusch RF (2015) Institutions and axioms: an extension and update of service-dominant logic. J Acad Mark Sci 44(5):20
Vink, J, Tronvoll B, Edvardsson B, Wetter-Edman K, Aguirre M (2017) Service ecosystem design: doing institutional work through design. In: Proceedings of the 2017 Naples forum on service, Naples, 6–9 June 2017

Chapter 7
Navigating Service Design Tools Using the Map of Capabilities

Abstract This is a *navigation* chapter, with the aim of providing a logical frame to summarise the design capabilities at all the levels described in the previous chapters. The reader can use the navigation framework suggested in this chapter to create their own *operational paradigm* or as a way to use the various tools and toolboxes available in service design literature.

The previous chapters provided an overview of design capability in relation to the levels of design intervention. The aim was to give an understanding of the value of service design and design action in the evolution of contemporary society and provide a view with different lenses—from the details of everyday life to the broader overview needed to understand and possibly orient societal change.

As a profession, service design today is becoming more and more relevant, and the offering for service design education (Becermen and Simeone 2019, 2020) and training tools[1] is increasing. However, the characterisation of the professional role of service designers in various aspects of society is nevertheless quite underdefined. Evidence of this is in how few calls for positions in the public or private sector mention service design as a specific professional profile. While companies or public institutions have a very clear idea of what a medical doctor, or an architect, or a lawyer can do—and they understand the specialisations within their professional area—the demand for service designers is *hidden* behind other job titles, such as *interaction designer*, *UX designer*, and *experience designer*. It is not that these titles are better defined, but they do have the advantage of being referred to in very specific phases of the development of a new product or service (Ehn et al. 2020).

A professional area is recognised once its practices are defined and implicitly or explicitly codified. And when a toolbox is defined, it specifically refers to those practices. The codification of practices is a long and complex process based on the exchange of academic knowledge and the transmission of such knowledge through

[1]New initiatives are emerging that provide training opportunities to include design as a diffuse capability. Among them, the Designscapes project is developing a series of training modules for Design-Enabled Innovation in the urban context, with the support of a critical overview of some key tools (Designscapes.eu).

N. Morelli et al., *Service Design Capabilities*, Springer Series in Design and Innovation 10, https://doi.org/10.1007/978-3-030-56282-3_7

education and training. This is in fact what the growing number of academic education programmes in service design are currently effecting.

Defining a practice for service design means developing an *operative paradigm* (Arbnor and Bjerke 1997)—a specific conceptual and operational toolbox for service designers—which includes tools, methods, examples, cases, techniques borrowed and adapted from other disciplines, and anything else that may constitute a service design expertise. This activity is mostly developed through academic or professional conferences and publications, and is facilitated by the increasing number of toolkits available in the relevant literature and online publications.

The abundance of methods and tools available for service designers may inspire a *cookbook* approach, as in, an approach in which, given the problem to solve, it is possible to find a *recipe* (a method or a tool) that supports designers in finding the solution to the problem. However, the reality is far from being that easy, for three main reasons:

1. The number of methods and tools available in the various sources is wide enough to confuse an inexperienced service designer or a student.
2. Although the utilisation of tools and methods concerns the way designers address operative aspects, the way such methods and tools are used strongly depends on the designer's methodological approach. This, in turn, depends on each individual designer's attitude towards the interpretation of reality (their *ultimate presumptions*) and on the influence of the socio-technical landscape in which the designer operates.
3. There is no direct or unequivocal correspondence between a problem to solve and the tools that can be used to solve it.

The first point would require each individual designer to acquire their own navigation experience among the tools and the knowledge available to face operative tasks. Today, a number of toolkits are currently available, some of which serve as a reference for students and service design practitioners (Stickdorn and Schneider 2011; Stickdorn et al. 2018; Servicedesigntools.org; Polaine and Løvlie 2013). Service design practice is also consolidated through the acquisition of methods, tools and expertise from other academic or practice-related areas, such as social studies, anthropology, engineering or management studies. Therefore, other toolkits have been used in service design practice that originate from social innovation (Kimbell 2013b), service innovation (Kimbell 2013a), or studies on design and creativity (Sanders and Stappers 2012). The point is that the existence of so many tools does not define per se a practice of service design, just as the existence of a hardware store does not necessarily imply the existence of a plumber or a locksmith.

The second point would require designers to define their conceptual position in respect to the tasks they are required to perform. The same problem could be solved through a massive use of technology or through an approach that privileges customers' engagement, participation or collective intelligence. It can be grounded on a collective initiative (bottom-up) or from decision-making centres (top-down). This is valid for broad social problems, such as the shift to a circular economy, and in minimal interactions, such as the way people interact with their bank. The two approaches would lead towards different actions, and consequently, to the use of different tools: a technology-oriented approach to regulate the interaction between banks and their clients would suggest a *prescriptive* attitude, in which the action of the various actors in the service is supposed to follow a procedure specified by a blueprint or by rigidly defined use cases. A *participatory* approach to the same kind of interaction would be based on providing as much information as possible to the customer in order to create some expectations that would support a more collaborative approach between the bank employee and the client.

The third point is a direct consequence of the first two—every problem can be seen from different perspectives and therefore can be solved through the use of different tools. And the same tool can be used to solve different problems in different phases of the design process.

7.1 Towards an Operative Paradigm

The definition of an *operative paradigm* is the process of the progressive accumulation of personal experience and knowledge that allows service designers to navigate between a number of problem areas using the most adequate tools and methods. Such tools and methods are often adapted from other disciplines and incorporated into the everyday practice of service designers. The various toolkits and design manuals available in the literature support the process of adapting different tools and methods to the process of service design. The navigation and use of such tools in design practice is instead related to the methodological approach, especially considering the operative tasks designers usually have in a design process. This book starts with the observation that the proliferation of toolkits is not accompanied by a corresponding proliferation of critical contributions to create an operative paradigm. To fill this gap, this book proposes a framework, starting from a classification of tools for service designers according to three main categories of tasks (Morelli and Tollestrup 2007):

- The analysis and interpretation of the context for the design action (analytical tools): in this area, tools are included that help the designer/design teams to understand the context of design action. This includes the identification of the main actors, the analysis of specific and personal profiles, and the identification of socio-technical conditions of the context.
- The development of new solutions (design tools): this area includes tools to not only articulate new solutions and define the relation between such solutions but also engage users and facilitate creative problem-solving processes.
- The representation and communication of the new solutions among the stakeholders in the value creation process (representation tools): this area includes tools to support not only a shared understanding of the roles and interaction in a service but also visions and the ability to mediate the collaboration among stakeholders.

Table 7.1 Overview of design capabilities and tasks at different logical levels

Levels of interventions	Category of tasks		
	Analysing	Designing	Representing
Service as interaction	**Addressing the context** Identifying stakeholders Identifying relevant issues Analysing complex contexts/routinary behaviours **Engaging stakeholders** Supporting conversation	**Controlling experiential aspects** Prototyping Experiments **Engaging stakeholders** Supporting participation **Modelling** Co-designing solutions Facilitating creative problem-solving	**Vision building** Inspiring participants Generating scenarios
Service as infrastructure	**Addressing the context** Analysing stakeholders' networks Analysing motivations	**Building logical architecture** Proposing service architectures **Open problem-solving** Creating platforms for interaction	**Vision building** Visualising organisational structures **Building logical architecture** Blueprinting services Ecosystems maps
Service as systemic institution	**Addressing the context** Understanding ecosystems and power relations	**Modelling** Proposing business models **Working across different logical levels** Articulating missions	**Vision building** Generating visions/scenarios

As mentioned, the use of such tools depends on the problem to solve, which will differ according to the scale of it, the interpretation of those who are addressing it, and the methodological approach of those who want to solve it. In particular, the levels defined in the previous chapter can help us identify specific tasks at each level (Table 7.1).

7.1.1 Service as Interaction: Analytical Capabilities

Addressing the context Identifying stakeholders and understanding their role in the context of the interaction (a single touchpoint or an event) is key here. This includes the analysis of stakeholders' competences and knowledge, and the main technological, cultural, political or natural issues that might shape the interaction. Certain tools can be used to address the context:

- Mapping tools such as ecosystem or stakeholders maps—these tools provide an overview of the main elements of the context and their interaction.
- Narrative tools such as journeys, which highlight practices and routines or personas, thus providing information about behaviours and preferences.

Engaging stakeholders This implies understanding the nature of the interaction stakeholders may have in value co-creation. When designers have direct interaction with the users/beneficiaries of a service (design activism, co-creation sessions) they engage people, facilitate their problem-solving activities, and inspire them by using tools or boundary objects. To engage stakeholders, certain tools and methods can be used:

- 5 Whys: Asking *why?*-questions to a person or a working group can be a way to discover the deep motivations for their behaviour.
- Card sorting: This method is useful in analysing the context, as it can prompt discussions of important issues that may reveal relevant aspects of the context. Cards can be created out of pictures of the context or by using symbols and metaphors.
- User stories: These engage participants in narrative exercises that help them describe situations, interactions, routines and emotions.

In many other cases the designer may not be able to interact directly with the service beneficiary but can still trigger or support the value creation process by generating tools for conversations that help beneficiaries analyse the context or frame the problem. Such tools may include these two useful tools:

- Issue cards: These act as a support to the analysis of complex contexts or everyday routines by breaking down areas of complexity into simpler issues that can be analysed and discussed (see Fig. 4.1).
- Maps and templates: These serve as frameworks that simplify the analysis of a context or an issue by helping people identify relevant actors (e.g. stakeholders' maps), interaction and value creations (e.g. value network maps), or routines (e.g. journey maps).

7.1.2 Service as Interaction: Design Capabilities

Controlling experiential aspects This implies the use of the following:

- Prototypes or pretotypes, that can simulate certain functional aspects of possible solutions to test how they will shape the experience of the service. The definition of a prototype may vary, ranging from a very sketchy simulation of specific aspects (e.g. *Mechanical Turks*, service acting, video sketches) to complex and interactive aggregations of service components.
- Experiments, which are often placed in specific contexts, such as the urban context and based on a direct interaction between designers and other actors (such as

citizens or service providers). Experiments will provoke or trigger conversations upon logical, ethical or experiential aspects of possible solutions (see Sect. 4.4.2).

Modelling This activity involves facilitating opportunities for co-creation, such as workshops, hackathons, co-design and co-creation sessions, and innovation jams. The tools and methods used in this case aim at supporting creativity, problem solving and meaning making, and include:

- Card sorting: Using cards to support discussions in a working group.
- Journey mapping: Engaging the working group in the construction of possible journeys, or a narrative that describes a solution in a way that can be later implemented.
- Prototyping: Generating rapid prototypes or models of possible solutions that can be used to enact a specific functionality and investigate its functional or emotional aspects (see Fig. 4.2).

7.1.3 Service as Interaction: Representation Capabilities

Vision building This capability is geared towards generating visions and highlighting perspectives that engage people into preferable scenarios and consequently orient present actions.

The role of designers in this case is to represent a future that is not yet visible. To accomplish this, designers can choose from certain narrative tools, such as:

- Tomorrow's headlines: creating the headline of a fictional newspaper article that describes a future event based on a possible scenario of the future.
- Journeys and storyboards: narrative, and often graphical, representation of a story in a possible/desirable future.
- Prototypes or video sketches: in addition to being a valid way for testing possible solutions, prototypes make it possible for people to imagine new lifestyles, new practices or new behaviours and therefore help align the visions and expectations of different actors.

7.1.4 Service as Infrastructure: Analytical Capabilities

Addressing the context This capability enables designers to analyse and interpret the ecosystem of the service that will support value creation. When working at the infrastructuring level, the analysis of the context needs to extend to stakeholders and contextual elements that may not be visible or relevant in single interactions. Here, as in the analysis of the level of 'service as interaction':

- Maps can be used to identify stakeholders and relevant components of the ecosystem, their role and their motivation.

- Journeys or other narrative tools can be used to figure out sequences of actions and stakeholders' behaviours.
- Personas can be used to analyse human, behavioural or social aspects that shape the context.

7.1.5 Service as Infrastructure: Design Capabilities

Building logical architecture This is the capability to create frameworks or logical and organisational structures that identify the main stakeholders and their role in shaping the service ecosystem. While the design capabilities to support services as interaction for value creation consist of proposing, provoking or inspiring, designing at this level implies a clear definition of the structures of interaction. It could have a prescriptive character, for example, when it describes the conditions for a function to happen, and it may be specific to the point of defining minimal aspects, such as the time or the channel of an interaction.

This capability implies analytical work on time sequences, service ecosystems and service platforms. Therefore, certain tools are often used to support this capability:

- Tools to control time, which include use cases, service journeys and service blueprints, which analyse the sequence of actions, specifying the actors and pre/post conditions for each interaction.
- Tools to control interaction, such as touchpoint mapping, which create maps of the different interaction between services and their stakeholders or beneficiaries.
- Tools to map the ecosystem (ecosystem maps), which include *synchronic* representations[2] of the actors in the ecosystem, including the flows of interaction (e.g. information or money exchange, physical movements) or indications about the actors' role, skills and contribution to the ecosystem.

Open problem solving Even when structures need to be created, the design of a service needs to take into consideration the fundamental principle that value is ultimately created by beneficiaries (Vargo and Lusch 2004, 2008). As a result, the architecture defined by a designer will be interpreted, used and possibly modified by those who will interact with the service. In some cases, infrastructuring a service means to create a space for conversation, collaboration or confrontation among different actors. This is the case, for instance, of hackathons or public innovation places, such as makers' spaces. In those cases, designers have to create frameworks that specify the rules for action and interaction, the roles of the actors, and the value that can be exchanged in addition to highlighting motivation and engagement issues.

---

[2]The term *synchronic representations* refers to a representation in which different interactions are synthesised regardless of their sequence in time. A map is a synchronic representation because a number of elements and their interactions are represented in one drawing. In contrast, *diachronic representations* are representations in which a time sequence is represented; a journey or a storyboard are examples of such a case.

Hackathons, co-creation workshops and design jams are the most common strategies to create spaces for mutual interaction. Such strategies involve very specific events in time that support communities in generating their own value, which designers can directly facilitate. However, open problem solving also concerns platform services, which create permanent or long-lasting physical or virtual spaces for value co-creation, such as that of service platforms. In these cases, certain tools are used to support open problem solving:

- Motivation matrix, which provides a structure of the motivations that would engage different stakeholders in the service. The matrix is a detailed description of the value each stakeholder would provide to the other stakeholders they will interact with in the service system.
- Platform canvasses, which can be used to organise the interaction among different actors in peer-to-peer interactions or in platform services.
- Business model or value proposition canvasses, which help designers specify how value is created, who contributes to it, which channels can be used, and what the financial streams are.

7.1.6 Service as Infrastructure: Representation Capabilities

Vision building This is the capability to visualise the logical and organisational infrastructure of a service or a platform by generating perspective views of how the service or platform will be used. In some cases, this capability requires the use of the same representation techniques used for generating architectures, taking into account that the aim of vision building is to communicate values and opportunities to the potential beneficiaries of the service. Therefore, vision building often requires visual or graphic representations to specify how the service structure will contribute to value creation, the possible scenarios of use and also the role, capabilities, and knowledge of each stakeholder.

Tools used for visualisation may include:

- Scenarios: meaning the definition of scenarios for future solutions on the basis of the projection of relevant critical factors. The projection of such scenarios on the present situation can be used to create organisational structures that take into account actors, resources, and motivations in the perspective of the desired future (see Fig. 7.1).
- Video sketching or other narrative techniques that make it possible to communicate aspects such as experiences, expectations or feelings that may not emerge from schematic visions of possible futures.

Common design tools, such as service journeys, can be used to provide future configurations of a service with a high communicative and inspirational power.

Building logical architecture The construction of logical or organisational architectures are often the outcome of a negotiation among different actors. Therefore, the

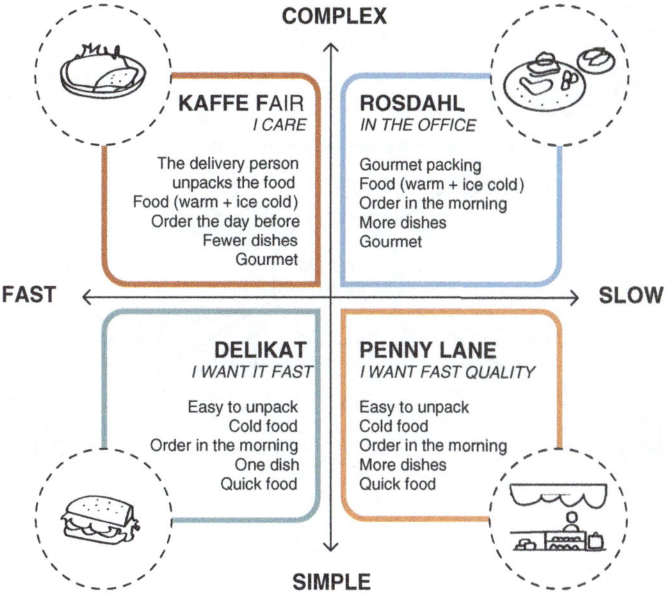

Fig. 7.1 Scenarios for different configurations of a lunch courier service. Adapted from Jonas et al. (2006)

representation of potential architectures for the service is important for the understanding of the stakeholders and for the negotiation of their role. The most common tools used to represent service architectures include:

- Time-related representations, such as service blueprints, that include the sequence of interactions and specify the actors and the channels of such interactions.
- Ecosystem mapping, which highlights the way stakeholders interact with the ecosystem or the benefits and contributions they would get from their participation in the service (Fig. 7.2).

7.1.7 Service as Systemic Institution: Analytical Capabilities

Addressing the context Working at the level of systemic institutions involves addressing extremely complex challenges, such as sustainability, social equality or policy-related issues, which are hard to describe or control. Addressing the context in this perspective means applying a systemic perspective and mapping systems with the aim of understanding, or at least making plausible, hypotheses about the way different actors and contextual factors are—or could be—directly or indirectly influencing the behaviour of the system. In addition to the analysis of human and non-human elements, the analysis of the context should also consider the strategies, policies and institutional arrangements that shape the ecosystem.

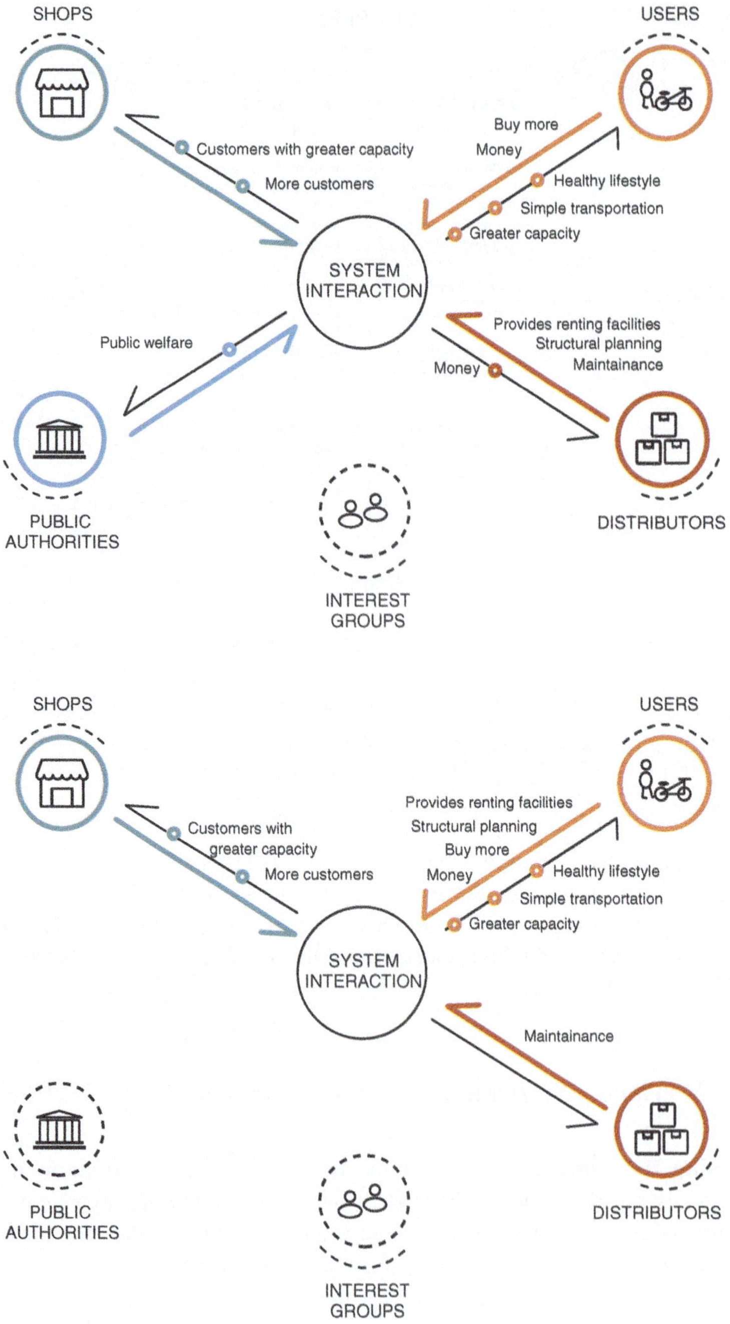

Fig. 7.2 Ecosystem mapping. Mapping the ecosystem can be used to visualise different scenarios/configurations of a service and make hypotheses on the role of each stakeholder. Adapted from Morelli (2006)

Fig. 7.3 Logical mapping of the context for the current ecological crisis. Adapted from https://commonstransition.org/toward-regenerative-society-plan-for-rapid-transition#prettyPhoto

The literature on design interventions to support change in systemic institutions does not specify the methods and tools that have been used to analyse a context. The relatively new role design plays in institutional changes calls for a better definition of specific tools. Systemic contexts, however, can be analysed by:

- System maps, which should include direct interactions between the main stakeholders and interactions with the legal, economic and technological factors. Several types of system analyses map the main issues and current trends and how they influence the present configuration of a system. These analyses are often supported by contributions from sociology, anthropology or economics (Fig. 7.3).
- Critical maps. Maps can also use analytical criteria as coordinates for understanding a systemic landscape and orienting the analysis of the present. The selection of critical factors, for instance, can generate different alternative interpretations of an existing context (Fig. 7.4).
- Moodboards can capture cultural vibes or social aspects that are hard to define in a quantitative description. Moodboards consist of a collection of images that convey general ideas or feelings that derive from a context (the 'character' of a city, or the attitudes of a community, or the *genius loci* of a place).

7.1.8 Service as Systemic Institution: Design Capabilities

Modelling The activity of modelling refers to simulating, visualising or experimenting possible solutions before all the information is available. The simulation of a new configuration of a systemic institution requires taking into account a high level of complexity, which is often hard to control. It is easier to imagine how an interaction can change among two actors in the same service, and it is relatively easier to understand how a service can change. However, it is very hard to imagine

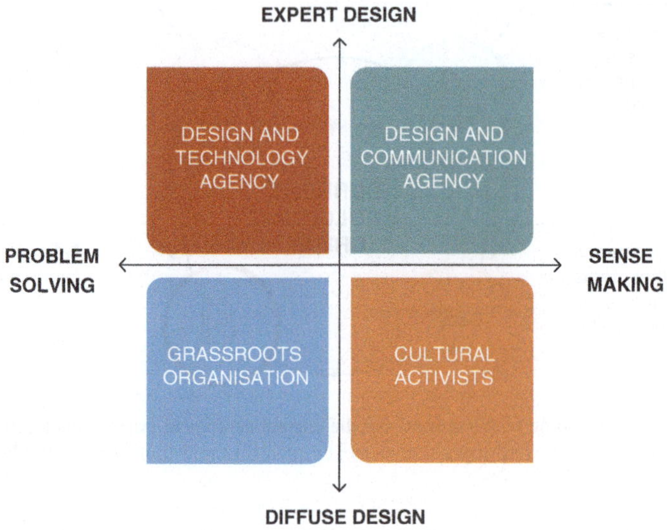

Fig. 7.4 Mapping design agency. Adapted from Manzini (2015)

how an institutional system (a political system, the healthcare system of a country, or cultural institutions) could operate a transition to a new configuration. Designers' capability to model solutions before all the information is available can be useful in those cases because it proposes possible aggregations of different critical factors. Modelling not only represents possible scenarios but also proposes how different actors can play a role using existing resources or contextual conditions within the perspective of desirable transitions.

Here again, tools or methods to support modelling capability at this level of complexity are not adequately documented in the literature because of the relatively new involvement of design as a discipline that deals with change on this scale. Other disciplines work on modelling by using canvasses, often derived or adapted from business model canvasses, to make their proposal for institutional change more concrete. Two tools in particular are often used in this type of modelling work:

- Business model canvases that have been proposed to control different aspects related to systemic changes, including value proposition, key partners, channels, costs and revenues. The business model canvas has been used to analyse business cases rather than systemic changes; however, the idea to use a canvas as a tool to *simplify* complexity inspired modified versions of the original tool. Canvases have been proposed to address changes in communities (Pfortmüller et al. 2017) or to create institutional infrastructures to involve citizens in the use of open data (Morelli 2018).
- Design-orienting scenarios consisting of the representations of possible and desirable futures in a way that can orient the present design decisions (Manzini et al. 2009). The scenarios are based on the analysis of systemic components and the

way they will recombine in future changes. Therefore, they offer a narrative view of the possible interplay between stakeholders, technologies, natural elements, social trends and uncertainties. Such scenarios are called 'design orienting' because they provide a framework to orient action in the present. Those visions of the future are therefore not simple projections of the present into the future but rather a critical activity of selecting among different possible futures in order to define preferable directions that can orient present action. They are led by motivations (where we want to go) and include practical indications about the next steps to take towards the preferred indications (Manzini et al. 2009).

Working across different logical levels Designing change in systemic institutions means figuring out broad changes in the social, technical, economic and political landscape. Imagining such change, and in general, figuring out the future has been an activity that has engaged experts in disciplines ranging from economics to literature. One specific contribution designers can offer is in the capability of translating abstract or narrative pictures of possible futures into operative indications on concrete action that also includes parameters for the evaluation of the impact of such actions.

Design capabilities should therefore concern the attitude to visualise possible directions, offering parameters for choosing those that are preferred (see the previous section on modelling) and the opposite attitude to project such long-term perspectives in the short-term future.

The complexity of the present ecosystems requires an active attitude towards the future that can be translated with the articulation of missions that can set the direction for production, distribution and consumption patterns (Mazzucato 2017). Design capability to work at different levels of abstraction could be used to translate such missions into action. A method that is often proposed for managing such a process is the Theory of Change.

The Theory of Change is a way of connecting a long-term transformation with the present by proposing possible 'change journeys' that specify outputs, outcomes and the impact of the proposed change, together with the action that happens in between. Theory of Change is a strategic planning tool that articulates and graphically illustrates the steps that need to be taken to realise a desired goal or impact and the expected results of these steps. It does so by starting from a set of assumptions and hypotheses about the present (theory of problem) and then shows the theorised causal pathways between a project's objectives, its activities, and its expected outcomes and impacts. It says, 'If we take action X, then this will cause effect Y, and this will eventually lead to outcome Z' (Simeone et al. 2019).

7.1.9 Service as Systemic Institution: Representation Capabilities

Vision building When visualising and representing large transformations of systemic institutions, designers are not the only ones who are able to contribute to generating tangible visions of what the future could look like. Literature, art and cinema are full of realistic representations of this kind. However, the role of designers in this case is to ensure that such visions are usable to orient change, meaning that they are coherent and include critical issues that can be addressed to orient short and long-term change. A purely fantastic vision or utopian/dystopian views of the future proposed by literature, art or cinema have been very effective in triggering debates about the directions in which we want to go, but they are not always useful when used to actively change systemic institutions by generating policies or long-term infrastructural change.

The previous sessions have already mentioned tools, such as the Theory of Change and design-orienting scenarios that can help visualise mission-oriented plans, highlight criteria or impact factors that would translate them into concrete actions.

7.2 References to the Tools Mentioned in This Book

The tools mentioned in this chapter and in those previous have been better explained in several toolkits, handbooks or academic papers. We created an additional table, which provides an overview of the main references, taking into account that the continuous production of knowledge on service design tools will likely offer new and more interesting sources. The table is located at this link (https://servicedesignlab.aau.dk/service-design-capabilities/) and will be further updated after the book has been published.

References

Arbnor I, Bjerke B (1997) Methodology for creating business knowledge, 2nd edn. Sage, Thousand Oaks and London

Becermen B, Simeone L (2019) Exploring the landscape of service design education: a preliminary review of current programmes in higher education. In: Proceedings of the international association of societies of design research conference, Manchester, 2–5 Sept 2019

Becermen B, Simeone L (2020) Current and future trajectories for service design education: views from educators in academia. In: Proceedings of ServDes2020, Melbourne, 2–5 Feb 2021

Ehn DE, de Götzen A, Simeone L, Morelli N (2020) Do they (know they) need a service designer? An investigation of service design capabilities through the lens of the market. In: Proceedings of ServDes2020, Melbourne, 2–5 Feb 2021

Jonas W, Morelli N, Munch J (2008) Designing a product service system in a social framework—methodological and ethical considerations. In: Proceedings of DRS conference, Sheffield, 16–19 July 2008

Kimbell L (2013a) The service innovation handbook: action-oriented creative thinking toolkit for service organizations. BIS Publishers, Amsterdam

Kimbell L (2013b) The social design methods menu. Retrieved from https://www.lucykimbell.com/stuff/Fieldstudio_SocialDesignMethodsMenu.pdf. Accessed 28 June 2020

Manzini E (2015) Design, when everybody designs. MIT Press, Cambridge

Manzini E, Jegou F, Meroni A (2009) Design oriented scenarios. Generating new shared version of sustainable product service systems. In: Crul M, Diehl JC, Ryan C (eds) Design for sustainability (D4S): a step-by-step approach. United Nations Environment Program, Delft

Mazzucato M (2017) Mission-oriented innovation policy: challenges and opportunities. UCL Institute for innovation and public purpose working paper (2017-1)

Morelli N (2006) The system around the product: definition of an operative paradigm for industrial design in a systemic context. In: Michael R (ed) Drawing new territories—best of design research. Swiss Design Network, Geneva

Morelli N (2018) Open4Citizens business models and sustainability plans. Open4Citizens project deliverable D4.10. https://open4citizens.eu/wp-content/uploads/2016/01/O4C_D4.10_30.06.2018_Final.pdf. Accessed 19 Apr 2020

Morelli N, Tollestrup CHT (2007) New representation techniques for designing in a systemic perspective. In: Proceedings of the second Nordic design research conference, Stockholm, 27–30 May 2007

Pfortmüller F, Luchsinger N, Mombartz S (2017) Community canvas guidebook. The guide to building meaningful communities. https://s3-us-west-2.amazonaws.com/openideo-resources/production/assets/1485.pdf. Accessed 23 June 2020

Polaine A, Løvlie L (2013) Service design: from insight to implementation. Rosenfeld Media, Brooklyn

Sanders EBN, Stappers PJ (2012) Convivial toolbox. BIS Publishers, Amsterdam

Simeone L, Drabble D, Iacopini G, Van Dam K, Morelli N, de Götzen A (2019) Articulating a strategic approach to face complexity in design projects. In: Proceedings of the academy for design innovation management conference, London, 18–21 June 2019. https://doi.org/10.33114/adim.2019.01.188

Stickdorn M, Schneider J (eds) (2011) This is service design thinking. BIS, Amsterdam

Stickdorn M, Lawrence A, Hormess M, Schneiderr J (2018) This is service design doing. Applying service design thinking in the real world. A practitioners' handbook, 2nd edn. O'Reilly, Sebastopol

Vargo SL, Lusch RF (2004) Evolving to a new dominant logic for marketing. J Mark 68:1–17

Vargo SL, Lusch RF (2008) Service-dominant logic: continuing the evolution. J Acad Mark Sci 36(1):1–10

Chapter 8
Beyond This Book

Abstract Some concluding notes originate from the view proposed by the previous chapters. This particular view is like a picture that captures a portion of a landscape in a specific frame, time and logical context. This chapter clarifies this frame and suggests ways to extend this view beyond the logical, professional and time limitations that a book could possibly present.

This book offers an overview of the capabilities that a service designer should acquire to be able to qualify their contribution to society. Many other possible topics have been left out of this book because, although very relevant, they are not specific to the profession of a service designer. Furthermore, the reader should take into account that a book is a picture of a context in time, where the context is not static and is extended much beyond the frame and the time of the picture. Certain considerations, which are outlined in the next section, position the book in respect to the professional, logical and historical context in which a service designer would operate.

Beyond service design competence This book is grounded on the belief that designers' capabilities need to be complemented with other professional capabilities that are essential for consolidating innovation processes. The collaboration between design and professionals from other disciplines, such as technical experts, anthropologists, sociologists or management experts, has been implicitly assumed in this book. Another relevant aspect that has only been marginally touched upon in this book is the evaluation of the impact of design action. The many toolkits and handbooks on how to design services does not correspond with any quality research on how such tools are used and the extent to which the use of such tools has been effective. However, this is hardly surprising, given that service design capabilities are generally not measured according to any quantitative parameters. Such a quantitative measurement could clearly validate the contribution of designers in innovation processes according to a logic that could help convince investors or the financial departments of public institutions. The parameters of such a quantitative measurement have never been defined and probably cannot be defined in absolute terms, although studies have been conducted that focus on the criteria and strategies for an evaluation of services from mixed quantitative and qualitative perspectives (Foglieni et al. 2013, 2018).

N. Morelli et al., *Service Design Capabilities*, Springer Series in Design and Innovation 10, https://doi.org/10.1007/978-3-030-56282-3_8

Beyond the frame As mentioned in the first sections, the aim of this book was not to create a new design manual for our students, as many and better manuals that offer detailed descriptions and examples of how service design tools are used are already available. Instead, the aim was to provide a navigation tool to read service design action in relation to the capabilities that are now required of service designers.

The authors discovered that although several tools are available for operating in the growing area of service design, some concepts and perspectives of this discipline, and on innovation in general, have not been properly integrated into a methodological approach. This encouraged the authors to attempt to see how the role and capabilities of service designers can be described in the light of two fundamental logical changes:

- The first change concerns the way to view value production in services, as introduced and explored by the service dominant logic. Such changes in fact shift the main role in value production from service providers (and a designer supporting them) to service beneficiaries. When looking at designers' capabilities and roles, this change prompts the question—*what is the role of designers, and what capabilities should they have in the service dominant logic?*
- A logical change deriving from a multi-level perspective that observes innovation paths at different scales. This perspective suggests the question—*how are service design capabilities used at different levels of innovation?* In this respect, this book proposes a navigation map that would help them understand which tool to use and how to use it according to the scale of change they are operating to.

This is the point of this book. And the authors believe this approach will also help support designers when navigating the complexity of their profession in light of the changes to come.

Beyond the (present) time This book is being written in an historical moment characterised by a complex and possibly revolutionary change that will leave a deep footprint on the way our society is organised.

In 'normal' times, the operative domain of design would be mostly limited to changes at the levels of 'service as interaction' and of 'service as infrastructure'. Most of the academic contributions on innovation processes are based on the assumption that the overall cultural, social, economic and political landscape in which our living systems are organised are changing at a very slow pace and according to evolutionary logics in which human beings, and therefore, designers, have very little control.

Many authoritative sources in innovation literature, however (Schumpeter 1943; Dosi 1982; Kuhn 1962) point out that the history of innovation is not a constant progress but rather proceeds by alternating long periods of slow development and short periods of revolutionary changes. The hypothesis that the COVID-19 crisis, and its planetary character, may be one of those revolutionary changes would need an in-depth analysis, which is not in the scope of this book. However, some hypotheses can be made on the nature of the coming change that have a much more operative character.

The crisis is prospecting a radical change in relationships between people, the organisation of services, and future technological, economic and environmental policies, and perhaps, in the long term, this change will also influence political systems. This change cannot be predicted, but we need to prepare our eyeglasses to properly observe and interpret them. Indeed, multifocal glasses will be needed for a close look at the changes in the way we live and also for a distanced look on the way we interpret and reorganise our world.

As for any pair of multifocal glasses, the structure of the lens makes it possible to focus on different distances and scales. In line with this, the logical structure of this book proposes a similar observation tool that allows for an interpretation and mapping of different levels of reality and suggests different capabilities and tools that make it possible to navigate in such a reality.

References

Dosi G (1982) Technological paradigms and technological trajectories: a suggested interpretation of the determinants and direction of technical change. Res Policy 11:147–163

Foglieni F, Villari B, Maffei S (2013) Evaluating services: an exploratory approach beyond service design. In: Proceedings of the 2nd Cambridge academic design management conference, Cambridge, 4–5 Sept 2013

Foglieni F, Villari B, Maffei S (2018) Designing better services: a strategic approach from design to evaluation. PoliMi Springer Brief, Milano

Kuhn TS (1962) The structure of scientific revolutions. University of Chicago Press, Chicago and London

Schumpeter JA (1943) Capitalism, socialism, and democracy. G. Allen & Unwin, London

The manufacturer's authorised representative in the EU is Springer
Nature Customer Service Centre GmbH, Europaplatz 3, 69115 Heidelberg,
Germany. If you have any concerns regarding our products, please
contact ProductSafety@springernature.com

Printed and bound by CPI Group (UK) Ltd, Croydon, CR0 4YY
30/09/2026
02225710-0003